Edwin J. Houston

Electrical Measurements

and other advanced primers of electricity

Edwin J. Houston

Electrical Measurements
and other advanced primers of electricity

ISBN/EAN: 9783337405687

Printed in Europe, USA, Canada, Australia, Japan

Cover: Foto ©berggeist007 / pixelio.de

More available books at **www.hansebooks.com**

ELECTRICAL MEASUREMENTS

AND OTHER

ADVANCED PRIMERS

OF ELECTRICITY.

BY

EDWIN J. HOUSTON, A. M.,

PROFESSOR OF NATURAL PHILOSOPHY AND PHYSICAL GEOGRAPHY IN
THE CENTRAL HIGH SCHOOL OF PHILADELPHIA ; PROFESSOR
OF PHYSICS IN THE FRANKLIN INSTITUTE OF PENNSYL-
VANIA ; ELECTRICIAN OF THE INTERNATIONAL
ELECTRICAL EXHIBITION, ETC.

AUTHOR OF

" A DICTIONARY OF ELECTRICAL WORDS, TERMS AND PHRASES, "
" ELEMENTS OF PHYSICAL GEOGRAPHY," ETC.

NEW YORK :
THE W. J. JOHNSTON COMPANY, LIMITED,
41 PARK ROW (TIMES BUILDING).

LONDON :
WHITTAKER & CO., PATERNOSTER SQUARE.
1893.

PREFACE.

In presenting to the public the volume entitled "Electrical Measurements and Other Advanced Primers of Electricity," the author again calls attention to the fact that these primers are in no sense to be regarded as revisions of the "Primers of Electricity," published in Philadelphia in 1884, during the International Electrical Exhibition.

The International Electrical Exhibition Primers, written during the early days of the Exhibition, were intended to explain merely the elementary principles of electricity to a public, that was then almost entirely ignorant of even the rudiments of the science.

As is well known, the times have greatly changed since 1884. Multitudinous commercial applications of electricity have rendered it no longer optional whether or no the general public shall be acquainted with the principles of electrical science. Such knowledge has become a necessary part of everyday business life.

Primers, therefore, based on the simple lines of those of 1884, would now occupy but a comparatively limited field, and the author has therefore entirely rewritten his earlier primers and greatly enlarged their scope.

In these days of voluminous electrical literature the student is often in doubt as to the best books with which to begin his studies. As an aid in this direction, and as a species of University Extension work, there has been placed at the end of each primer extracts from one or more standard electrical books, so as to give the student some idea of their character, and thus enable him to intelligently select the works best suited to his needs.

The author desires to acknowledge his indebtedness to Mr. T. C. Martin and Mr. Joseph Wetzler for critical revision of some of the chapters.

EDWIN J. HOUSTON.

CENTRAL HIGH SCHOOL,
      Philadelphia, Pa.,
          January, 1893.

CONTENTS.

I.—THE MEASUREMENT OF ELECTRIC CURRENTS.

There are various methods for measuring the current that passes in any circuit. These methods are based on the electrolytic, the heating, or the magnetic power of the current. They may, therefore, be arranged under the following heads; namely:

(1.) The voltametric method, or by the use of voltameters, based on the electrolytic power of the current.

(2.) The calorimetric method, or by the use of apparatus called calorimeters, based on the heat produced by the current.

(3.) The magnetic method, or by the use of various apparatus called galvanometers, etc., based on the deflections of a readily movable magnetic needle, core, or movable circuit by the magnetic field produced by the current.

(4.) The indirect method, or that in which the values of the electromotive force and the resistance are obtained, and that of the current strength calculated therefrom by the well known formula:

$$C = \frac{E}{R}.$$

In the voltametric method an instrument called a voltameter is employed, the operation of which is dependent on the fact that the strength of current flowing in any circuit, or more correctly the quantity of electricity or the number of coulombs that pass per second through such circuit, can be determined from the amount of chemical decomposition effected.

Various chemicals are employed for such purposes ; the principal of these are dilute sulphuric acid or solutions of copper sulphate or of silver nitrate.

It can be shown that an electric current equal to one coulomb per second, or one ampère, will deposit .00111815 gramme, or .01725 grain of silver per second, or will decompose .00009326 gramme, or .001439 grain of dilute sulphuric acid of a specific gravity of about 1.1 per second.

Voltameters may be divided into two general classes ; namely, volume voltameters and weight voltameters. In the first the quantity of electricity passing is determined by the volume of gas evolved ; in the second by the weight of material decomposed after the current has passed for a given time.

In the sulphuric acid voltameter shown in Fig. 1,

electrodes of platinum are immersed in sulphuric acid of the specific gravity of 1.1. On the passage of the current hydrogen appears at the negative terminal and oxygen at the positive terminal, in the proportion of about two volumes to one. The electromotive force of the battery must not be less than 1.44 volts, or electrolysis will not occur.

Fig. 1.—A Sulphuric Acid Voltameter.

The sulphuric acid voltameter may be arranged either as a volume or as a weight voltameter. In the form shown in Fig. 1, it is arranged as a volume voltameter. When arranged as a weight voltameter, the evolved gas escapes through a drying tube containing calcium chloride. This tube is provided to stop the acid liquor that may be mechanically carried over with the disengaged gases.

The weight of sulphuric acid decomposed is determined from the decrease in weight of the instrument after the current has been passed for a given time.

The form generally given to the weight voltameter is that in which the current is passed between plates of copper immersed in an electrolyte of copper sulphate, or of silver immersed in silver nitrate. In either case the amount of current passing is determined by the increased weight of one of the plates.

If several voltameters of different sizes and shapes are placed in a circuit in series, and the same kind of liquid is placed in each instrument, such, for example, as copper sulphate, and a current is passed successively through them, it will be found that the amount of deposition as determined by the increase of copper on one of the plates is the same in each case. In other words, the amount of chemical decomposition effected is independent of the size, shape or construction of the voltameter, and depends only on the strength of the current that passes. This fact renders the voltameter a ready, though not very reliable, means for determining the strength of a current.

To determine accurately, however, the current strength passing by such means, a fairly considerable time is necessary, and during this time the current strength is exceedingly apt to vary. For this reason other methods are generally adopted in practice for the measurement of current strength.

The calorimetric method is based on the fact that the strength of current passing in any circuit may be determined from the energy liberated in such circuit as indicated by means of the increase in temperature produced in the cir-

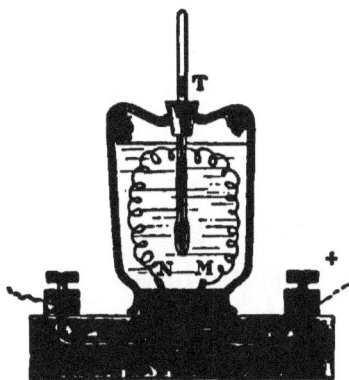

FIG. 2.—ELECTRIC CALORIMETER.

cuit. This increase in temperature is determined by means of an instrument called an electric calorimeter. It consists, as shown in Fig. 2, of a vessel provided with a liquid which surrounds a portion of the circuit $N\ M$, immersed therein. A thermometer, T, is provided for measuring the increase in temperature.

The indications of the calorimeter are based on the increase in temperature in a given time of a weighed quantity of water or other liquid in the calorimeter. This increase is proportional :

(1.) To the resistance of the conductor.

(2.) To the square of the strength of the current passing.

(3.) To the time the current is passing.

In the galvanometric method the strength of the current passing is determined by the use of certain instruments called galvanometers. This method is generally employed in commercial determinations of current strength, because such strength can be thus determined by a single observation.

A galvanometer is an instrument for measuring the strength of an electric current by means of the deflection of a magnetic needle. The galvanometer was invented by Schweigger, and is based on the discovery by Oersted of the power which an electric current possesses of deflecting a magnetic needle placed near it.

This deflection, as we have already seen, is due to the mutual action which exists between the field of the magnet and the field of the current.

The following principles should be borne in mind in the study of the galvanometer :

(1.) The direction of the deflection of the mag‐ netic needle will depend both on the direction in which the current flows in the deflecting circuit and on the relative positions of the circuit and the needle to each other.

(2.) No matter in what direction the needle is deflected it will, if the current is sufficiently strong,

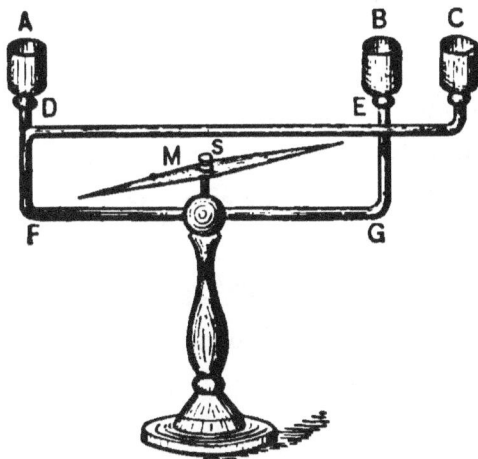

FIG. 3.—AMPÈRE'S APPARATUS.

always tend to come to rest at right angles to the circuit.

(3.) In the same instrument the amount of de‐ flection increases with the strength of the current that passes, though not necessarily proportionally.

The power possessed by an electric current of de‐ flecting a magnetic needle can be readily shown by means of the apparatus represented in Fig. 3, in which a conductor, $D\ F\ G\ E$, bent in the form of a hollow rectangle, is provided with a readily movable magnetic needle M, supported at its centre. Mercury cups A, B and C, serve to pass the current in different directions through the apparatus. In experimenting with this apparatus it will be found that the direction in which the needle is deflected will depend on the direction in which the current flows.

Various methods have been suggested for readily remembering the direction of deflection of a magnetic needle under various circumstances ; for example :

(1.) The north pole of a magnetic needle is deflected to the left hand of an observer who is supposed to be swimming in the current and facing the needle.

(2.) If an ordinary corkscrew, placed along a conductor through which a current is passing, be twisted so as to advance or move in the same direction as the current, the direction in which its handle must be turned in order to produce such motion is the same as the direction in which the needle will be deflected.

Since in all cases the needle comes to rest with the lines of force of the deflecting field passing in at its south pole and coming out at its north pole, and since the lines of magnetic force of an electric circuit form concentric circular lines around the circuit, the needle, as can readily be seen, must come to rest with its north pole pointing in diametrically opposite directions on opposite sides of the conductor.

By referring again to Fig. 3, and remembering either of the above rules as to the direction of the deflection of the needle, it will be seen that when the current passes through the rectangular circuit *D E G F*, each branch *D E, E G, G F,* and *F D*, will deflect the north pole of the magnetic needle *M*, in the same direction ; for, if the current passes above the needle through the branch *D E,* from left to right, it will pass through the lower branch, *G F,* from right to left, and will, therefore, deflect the needle in the same direction. So also the current passes through the vertical branch *E G,* in the opposite direction to that in which it passes through *F D,* and these two will also tend to deflect the needle in the same direction. Moreover, all the separate branches tend to deflect the needle in the same direction.

Based on this principle, Schweigger constructed a piece of apparatus called the multiplier, in which the deflecting current is caused to pass a number of times around a coil of insulated wire, *C C,* Fig. 4, wrapped around a hollow rectangular frame as shown. When a current is passed through the coil by connecting its terminals, *N, P,* to an electric source, the needle will be deflected to an extent that will increase with the number of turns of wire that are placed in the coil *C C,* provided, of course, the current strength remains constant.

FIG. 4.—SCHWEIGGER'S MULTIPLIER.

The galvanometer is based on Schweigger's multiplier. It consists of coils of various shapes and sizes so placed, as regards a readily movable magnetic needle, as to cause its deflection on the passage of the current. The current strength is determined by the value of such deflections, and this value will vary according to the size and number of the deflecting coils, the relative size of the needle, and its position as regards the deflecting coils.

In all cases, when no current is passing, the needle, when at rest, should occupy a position parallel to the plane of the coil. On the passage of the current the needle tends to place itself at right angles to the direction of the current, or to the plane of the conducting wire in the coil. The

FIG. 5.—ASTATIC GALVANOMETER.

needle is deflected by the current from its position of rest either in the earth's field or in the field obtained from permanent or electro-magnets. In the first case, when in use to measure a current, the plane of the galvanometer coils must coincide with the plane of the magnetic meridian of the place. In the other case, the needle may be used in any position in which it is free to move.

Galvanometers are constructed in a variety of forms, either according to the purpose for which they are employed, or the manner in which their deflections are valued. A very common form given to the galvanometer is shown in Fig. 5. This form is called the astatic galvanometer because it employs an astatic needle.

The astatic needle consists, as shown in Fig. 6, of two magnetic needles, $N\,S$ and $S'\,N'$, placed verti-

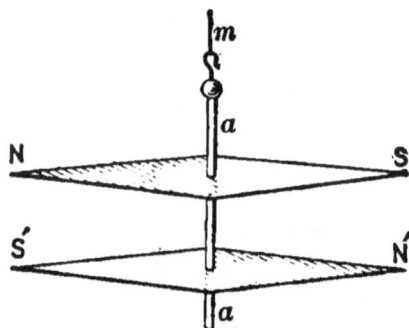

FIG. 6.—ASTATIC NEEDLE.

cally one above the other and rigidly connected to the vertical axis $a\,a$, with their opposite poles, $N\,S$ and $S'\,N'$, opposed to each other.

The idea of employing an astatic needle in this form of galvanometer is to increase the sensitiveness of the instrument, and thus obtain a greater deflection by the use of a smaller current.

When an astatic needle is employed, the upper

needle is placed outside the deflecting coil, and the lower needle inside the coil, as shown in Fig. 7.

Since the current passes under the needle $S N$, in the opposite direction to that in which it passes over it, both of these portions of the circuit deflect the needle in the same direction. The upper needle, $S' N'$, is deflected by the portion of the current flowing below it in the same direction as the lower needle, $N S$, because its poles are oppositely directed.

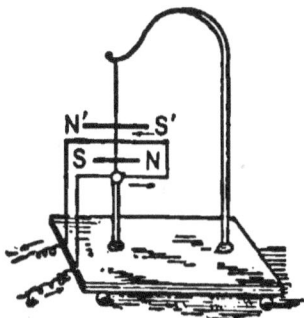

FIG. 7.—ASTATIC NEEDLE AND DEFLECTING CIRCUIT.

In very sensitive galvanometers two coils are employed, the one surrounding the lower needle and the other surrounding the upper needle. These coils are so wound that their deflecting action on both needles is in the same direction. In some forms of galvanometers the sensitiveness of the instrument is varied by means of a magnet called a compensating magnet, placed on an axis above the magnetic needle. As the compensating magnet is moved toward or from the needle, the effects of the earth's field, and, consequently, the sensitiveness of the galvanometer, are varied.

Galvanometers for commercial use assume a variety of forms. Their scales are generally so divided as to enable the ampères, volts, ohms, or watts, etc., to be read off directly. Such instruments are called ampère-meters or ammeters, voltmeters, ohm-meters, watt-meters, etc.

By the sensibility of a galvanometer is meant the readiness with which its needle is deflected by the passage of a current, as well as the extent of such deflection. The reciprocal of the current required to produce a deflection of one degree is called the figure of merit of the galvanometer. The smaller the current required to produce this deflection the greater is the figure of merit, and, consequently, the greater is the sensitiveness of the galvanometer.

In order to increase the effective length of the needle without increasing its actual length, the expedient is sometimes adopted of reading the deflections of the needle, not by the motion of the needle itself, but by the motion of a spot of light reflected from a mirror attached to the suspension system of the needle, so as to move the spot of light over a distant scale.

A well known form of mirror galvanometer, devised by Sir William Thomson, is shown in Fig. 8. The needle is attached to the back of a light silvered concave mirror, suspended by means of a single silk fibre, and is placed inside a coil of insulated wire. In some mirror galvanometers the mirror is attached directly to the suspension fibre.

FIG. 8.—MIRROR GALVANOMETER.

The compensating magnet $N S$, is used to vary the sensitiveness of the instrument.

The lamp L, is placed at the back of a slot in a white screen, on the other face of which is placed a graduated scale K. The light which passes through this slot is reflected from the mirror as a bright spot of light which is caused to fall on the graduated scale.

In Fig. 9 the details of the slot and back of screen are shown.

A form of galvanometer called the sine galvanometer is shown in Fig..10.

This galvanometer differs from other galvanometers in the fact that its vertical coil, *M*, is movable about a vertical axis, so that the coil can readily be made to follow the needle in its deflections, and so be kept parallel to it. The needle is placed inside the coil, and can be of any length smaller than

FIG. 9 —DETAILS OF GALVANOMETER LAMP AND SCALE.

the diameter of the coil. Its deflections are marked on the horizontal graduated scale at *N*.

When used to measure a current, the vertical coil *M*, is moved on a vertical axis over the horizontal graduated circle *H*, so as to keep the coil parallel with the needle. This vertical coil *M*, is moved until the needle shows no further deflection, though the coil is parallel with the axis of the needle. The strength of the current is then inferred from the amount of movement which has been given to the vertical coil over the horizontal circle *H ;* or, what is the same thing, by the distance the needle has

FIG. 10.—SINE GALVANOMETER.

been deflected from its original position of rest in the earth's field.

In the sine galvanometer the current strength is proportional to the sine of the angle through which the magnetic needle is deflected. This angle is most conveniently measured on the horizontal graduated circle *II*, as the angle through which it is necessary to move the coil *M*, from its position when the needle is at rest in the plane of the earth's meridian, to the position in which the needle is no longer deflected by the current passing through its coils, although they are still parallel to the needle.

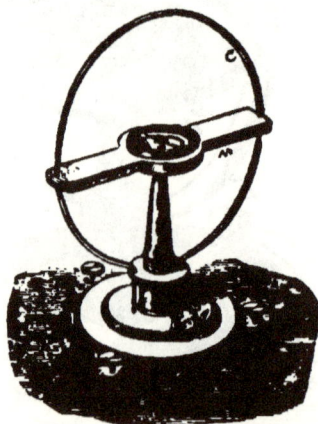

FIG. 11.—TANGENT GALVANOMETER.

In the tangent galvanometer the strength of current passing through the galvanometer coils is proportional to the tangent of the angle through which the needle is deflected. In this form of galvanometer the coil is fixed and the strength of the current passing is proportional to the tangent of the angle of deflection, provided the length of the magnetic needle is less than one-twelfth the diameter of the coil.

A form of tangent galvanometer is shown in Fig. 11. Tangent galvanometers, when used to measure large currents, consist often of but a single turn of wire. They are, however, frequently formed of a number of turns like other galvanometers. Great care must be taken to suspend the needle at the exact centre of the deflecting coil, and not to permit its length to exceed the above mentioned proportions.

It is often necessary to make galvanometric measurements on shipboard. Instruments for such purposes are generally termed marine galvanometers. In order to avoid the disturbing action caused by the rolling of the ship, the magnetic needle is suspended by a silk fibre attached above and below in a vertical line with the centre of gravity of the needle. It is also necessary to protect the needle from the influence of magnetized masses of iron in motion. To effect this the needle of such galvanometer is shielded from extraneous magnetic fields by means of a magnetic screen, which consists essentially of an iron box within which the whole galvanometer is placed.

In the differential galvanometer the needle is deflected by two coils of wire that are so wound and

placed as to produce deflections in opposite directions. The needle is, therefore, deflected to an
amount equal to the difference of the deflecting
forces.

FIG. 12. – DIFFERENTIAL GALVANOMETER.

When currents of equal strength flow through
each of the coils no deflection of the needle takes
place, since each coil neutralizes the other's effects,
so far as its action on the needle is concerned.

One form of differential galvanometer is shown in

Fig. 13. In some cases the separate coils may be so connected that each tends to deflect the needle in the same direction. In such cases, of course, the galvanometer ceases to be differential in action.

The form of galvanometer shown in Fig. 13 is called, after its inventors, the Deprez-D'Arsonval galvanometer. This form of galvanometer is also

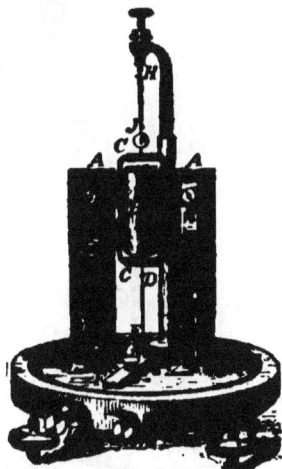

FIG. 13.—DEPREZ-D'ARSONVAL GALVANOMETER.

called a dead-beat galvanometer, from the fact that its needle moves rapidly over the scale to its position, and comes to rest without moving alternately on either side of its position of rest for a number of times, as is common in most forms of instruments.

The movable part of the Deprez-D'Arsonval galvanometer consists of a light rectangular coil *C*, formed of many turns of wire supported by two single wires, *HJ* and *DE*, between the poles of a strong, permanent horseshoe magnet *A A*. The dead-beat action is due to the fact that the motions of the coil under the action of the deflecting current

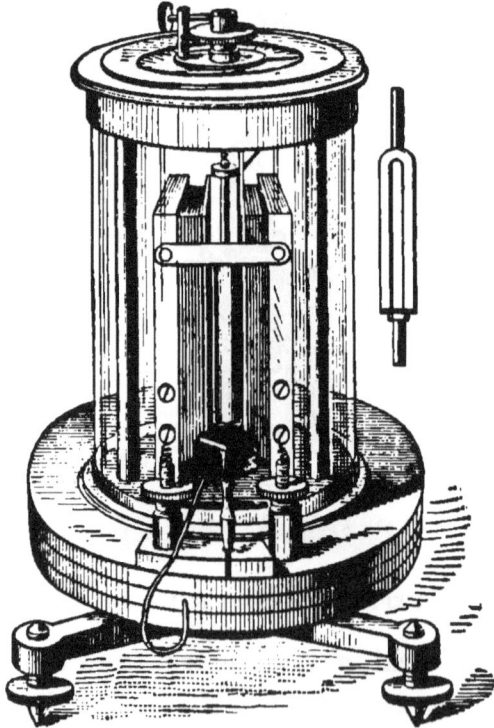

FIG. 14.—TORSION GALVANOMETER.

produce a current that tends to oppose such motion. The movements of the coil are observed by means of a spot of light reflected from the mirror *J*, fixed to the wire *HJ*.

In the torsion galvanometer the strength of the deflecting current is measured by the torsion it exerts on the suspension system.

In one form of torsion galvanometer the magnetic needle consists of a bell-shaped magnet, suspended by a thread and spiral spring between two deflecting coils of insulated wire, placed on either side of a magnet, as shown in Fig. 14.

FIG. 15.—VERTICAL GALVANOMETER.

The strength of the current to be measured is determined by the amount of torsion required to bring the magnetic needle back to its zero position, while under the deflecting power of the current. A horizontal scale is placed on the top of the instrument for accurately measuring the angle of torsion.

In the case of currents which continue for but a moment in one direction, for example, as that produced by the discharge of a condenser, a form of galvanometer known as the ballistic galvanometer is employed.

Galvanometers are sometimes divided into horizontal and vertical galvanometers, according to the position in which their needles are free to move. A vertical galvanometer is shown in Fig. 15. In such

FIG. 16.—DETECTOR GALVANOMETER.

an instrument the north pole of the needle is weighted, so that when no current is passing through the coils the needle points vertically downward to the zero on the scale.

Most galvanometers are provided with horizontal needles. In the form shown in Fig. 16, which is called a detector galvanometer, the needle is horizontal. Such an instrument is suitable for readily detecting the presence of a current in any circuit.

The form of galvanometer shown in Fig. 17 is known as Siemens' electro-dynamometer, and is suitable for the measurement of commercial currents.

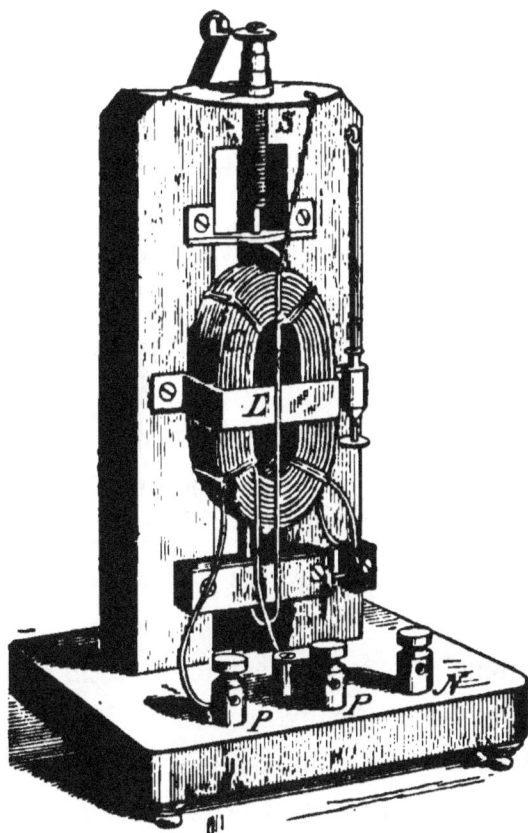

FIG. 17.—SIEMENS' ELECTRO-DYNAMOMETER.

The electro-dynamometer contains two coils, the fixed coil C, secured as shown to the upright support, and the movable coil L, consisting of but a single turn of wire. The movable coil is suspended by means of a thread and a delicate spring S, capable of being twisted through an angle of torsion that is measured on the horizontal scale shown at the top of the figure.

The ends of the movable coil dip into mercury cups, so placed in the circuit that the current to be measured passes in series through the fixed and the movable coils.

When ready for use the movable coil is placed at right angles to the fixed coil. On the passage of the current to be measured, the mutual action which these coils exert on each other tends to place the movable coil parallel to the fixed coil, against the torsion of the spring S. The amount of the deflecting force is determined by the amount of torsion required to bring the movable coil back to its zero position.

In Siemens' electro-dynamometer, unlike the torsion galvanometer already described, since the action of the fixed and movable coils is mutual, the current is proportional to the square root of the angle of torsion, and not, as in the torsion galvanometer, to the angle of torsion. Or, in other words, in the electro-dynamometer the deflecting force is proportional to the square of the deflecting current strength, while in the torsion galvanometer it is merely proportional to the current strength.

In the commercial distribution of electricity for the various purposes of light and power some form of instrument is required for measuring and recording the current which is supplied to each consumer ; or, more correctly, the quantity of electricity that passes in a given time through any consumption circuit. The purpose of such apparatus is the same as that of the meters employed to determine the quantity of illuminating gas consumed. They are for this reason generally called electric meters.

Electric meters are constructed in a great variety of forms; these, however, may be divided as follows, namely :

(1.) Electro-magnetic meters, or those in which the current passing is measured by the magnetic effects it produces.

In electro-magnetic meters the entire current may be passed through the meter.

(2.) Electro-chemical meters, or those in which the current passing is measured by the electrolytic decomposition which it effects.

In electro-chemical meters a known fractional or shunted portion of the current is passed through a solution of a metallic salt, and the current strength is determined by the amount of electrolytic decomposition effected.

(3.) Electro-thermal meters, or those in which the current passing is measured by movements effected by the increase of temperature of a resistance through which the current passes, or by the difference of weight produced by the evaporation of a liquid by means of heat generated by the current.

(4.) Electric time meters, or those in which no attempt is made to measure the current passing, but in which a record is kept of the number of hours that the current passes through the consumption circuit.

Edison's electric meter is of the second class. It consists of two voltameters formed of plates of zinc dipped in a solution of zinc sulphate. These plates are weighed at stated intervals—one every month, and the other every three months.

EXTRACTS FROM STANDARD WORKS.

Concerning the relative advantages of voltameters and galvanometers in the measurement of electrical currents, Ayrton, in his " Practical Electricity," * speaks thus on page 20 :

The disadvantage of employing a voltameter for the practical measurement of currents is that it requires a strong current to produce any visible decomposition in a reasonable time. Even the current of one ampère, which is about that used in an ordinary Swan incandescent lamp, would require two hours fifty-eight minutes and forty five seconds to decompose one gramme of dilute sulphuric acid, whereas the weak currents used in telegraphy, and, still more, the far weaker currents used in testing the insulating character of specimens of gutta-percha, india-rubber, etc., might pass for many days through a sulphuric acid voltameter before their presence could be detected, much less their strength measured. Indeed, not to mention the enormous waste of time, and the difficulty of keeping the current strength which it was desired to measure constant all this time, the leakage of gas which would take place at all parts of the apparatus that were not hermetically sealed

---

*"Practical Electricity : A Laboratory and Lecture Course for First Year Students of Electrical Engineering, by W. E. Ayrton, F.R.S. London: Cassell & Co., Ld. 1888. 516 pages, 173 illustrations. Price $2.50.

would render such a mode of testing quite futile. Hence, although the voltametric method is the most direct way of measuring a current strength, and although it is constantly made use of for measuring the large currents now used industrially, still the very fact that the amount of chemical decomposition produced in a given time by a certain current is independent of the shape or size of the instrument makes it impossible to increase its sensibility. Consequently some other apparatus must be employed for practically measuring small currents, and the law of the apparatus, that is, the connection between the real strength of the current and the effect produced in the apparatus, must be experimentally ascertained by direct comparison with a voltameter.

But if we are going to compare together the indications of the two instruments produced by various currents, the second instrument cannot be much more sensitive than the voltameter, and what advantage can arise from employing such an instrument? This leads us to the fact that, whereas in a voltameter there is only one way by which the production of the gas can be more easily measured, namely, by diminishing the bore of the graduated tube t (Fig. 5), up which the liquid is forced by the production of the gas, there are two quite distinct ways in which the magnitude of the deflection of a "*galvanometer*" needle can be more easily read. The first consists in using a microscope or some magnifying arrangement, or in simply lengthening the pointer, both of which methods correspond with using a tube of smaller bore in a voltameter; the second consists in winding a long fine wire, instead of a shorter thicker wire, on the bobbin of the galvanometer, and which causes the deflection of the magnet to be greater with the same current. This second mode has no analogy with any possible change in a single voltameter.

Now experiment shows *that a galvanometer of a particular shape and size, and with a definite magnetic needle, acted on by a definite controlling force, produced, say, by the earth's magnetism, or by some fixed permanent magnet, has a perfectly definite law connecting the magnitude of the deflection with the strength of the current producing it,* although the absolute value of the current in ampères necessary to produce any particular deflection can be increased or diminished by using fewer turns of thick wire or more turns of fine wire to make a coil of the same dimension.

Bottone, in his book, " Electrical Instrument Making for Amateurs,"* thus describes, on page 134, the making and mounting of a needle for a tangent galvanometer:

The tangent galvanometer presents no difficulty in construction. A small lozenge-shaped "needle" is made from a thin piece of watch spring, about 1 in. long and $\frac{1}{4}$ in. wide. This is "let down," or softened, by being held over the flame of a spirit lamp until of a dull red, and allowed to cool gradually. When quite cold a small hole $\frac{1}{8}$ in. in diameter is drilled through the centre. The "needle" is then straightened out, and tested for centrality; and, if

---

*"Electrical Instrument Making for Amateurs : A Practical Handbook," by S. R. Bottone. Fourth edition. London: Whittaker & Co. 1889. 202 pages, 59 illustrations. Price 50 cents.

defective, filed until the hole corresponds with the centre of gravity. It is then *hardened* by being made red hot over the flame of a spirit lamp, and being dropped into cold water. It must then be carefully magnetized by being rubbed at each extremity with the opposite poles of a good horseshoe magnet. When fully magnetized it must be fitted with a small glass pivot, made as described at paragraph six, small enough to enter the $\frac{1}{16}$ in. hole in the needle, and about $\frac{1}{4}$ in in length. Great care must be exercised in the choice of a pivot, which must be very perfectly shaped, so as to allow great freedom of motion in the poised needle. This point being settled, the pivot is attached to the needle by means of a mere trace of good glue, applied to the hole in the needle only. The needle must now be poised by its pivot on a fine steel sewing needle (No. 10 will do), and any want of perfect horizontality must be remedied while the glue is still moist. When the above is quite dry, a very fine straw, about $2\frac{1}{4}$ in. long, has a small hole made in its centre (half way between its two extremities) with a rather coarse pin ; then the head of the pivot is pushed through this hole in the straw, so as to cause the straw to lie exactly at right angles over the needle. The merest trace of glue will now cause the straw to adhere to and retain its position on the glass pivot. This can now be set aside to dry.

II.—THE MEASUREMENT OF ELECTRO-MOTIVE FORCE.

The electromotive force of a source, or the differ-ence of potential between any two points in a cir-cuit, can be measured in various ways by the use of instruments called voltmeters. Among the most important of these methods are the following :

(1.) By the use of galvanometers, or by galva-nometer voltmeters.

(2.) By the use of electrometers, or by electrom-eter voltmeters.

(3.) By the method of weighing, or by balance-voltmeters.

In the galvanometric method, differences of po-tential are determined by the quantity of electricity that flows per second through a given resistance, just as the pressure of water at any opening in a vessel can be determined by the quantity of water that flows out from such opening per second. Differences of potential, therefore, may be measured by means of any galvanometer which measures the current in ampères, provided the resistance of the circuit is known. Galvanometers constructed so as to meas-

(39)

ure differences of potential are called voltmeters, or, more correctly, galvanometer-voltmeters.

In the electrometer-voltmeter the difference of potential may be employed to charge insulated conductors, and the value of such differences of potential determined from the electrostatic attractions and repulsions acting on a readily movable needle suitably suspended near such charged conductors. This latter form of voltmeter is generally termed an electrometer, or an electrometer-voltmeter.

This method consists in ascertaining the force or weight required to overcome the attraction between two oppositely charged plates, or two oppositely energized coils, or by measuring the repulsion between two similarly energized coils.

When a galvanometer is used as an ampère-meter, for determining the strength of the current passing, it is placed directly in the main circuit. When used as a voltmeter, for determining the difference of potential between any points in the circuit, it is placed in a shunt circuit to these points. The coils of voltmeters are generally made of much higher resistance than those of ampère-meters.

According to Ohm's law :

$$C = \frac{E}{R}$$

Therefore : $E = C R$;

or, in other words, the electromotive force, or difference of potential in volts, is equal to the current in ampères multiplied by the resistance of the circuit in ohms.

Galvanometer-voltmeters may be constructed in a great variety of forms. In all such forms the difference of potential is determined from the deflection of a magnetic needle by the magnetic field produced by a current which flows through a coil of insulated wire. Since the resistance of the voltmeter is constant, the current passing, and hence the deflection of the needle, will vary only with the electromotive force.

In galvanometer-voltmeters the magnetic field produced by the current may deflect the needle of the galvanometer—

(1.) Against the earth's field.

(2.) Against the field of a permanent magnet or an electro-magnet.

(3.) Against the action of a spring.

(4.) Against the force of gravity acting on a weight.

Instead of determining the difference of potential by varying the magnetic field of the current produced, such current may be used to heat a wire, and the value of the current strength, and, consequently, the difference of potential, determined by the movement of a needle caused by the expansion of a wire.

There are, therefore, various forms which may be given to voltmeters, and various principles by which they operate.

A form of galvanometer-voltmeter devised by Sir William Thomson is shown in Fig. 17. A coil of insulated wire *A*, whose resistance is over 5,000 ohms, acts on a needle formed of a number of short

FIG. 18.—GALVANOMETER-VOLTMETER.

parallel needles placed one above the other. The compound needle so formed has attached to it a light aluminium index moving over a graduated scale.

A small circular magnet *B*, called the restoring magnet, placed over the needle, is used to vary the strength of the earth's field at any place, either by moving the magnet itself, or by moving the box containing the compound magnetic needle toward or from the deflecting coil. The indications of this instrument are based on the fact that when a galvanometer of sufficiently high resistance is connected with any two points in a circuit, the current which passes through it, and, consequently, the deflection of its needle, is directly proportional to the difference of potential between two such points.

It is not necessary, in measuring the difference of potential between any two points in a circuit or conductor, that such differences of potential be utilized to cause an electric current; they may, instead, be used to produce charges in an insulated conductor, and the differences of potential can then be inferred from the movements of a needle as the result of electrostatic attractions and repulsions produced by such charges; or they may be determined from the weight required to balance such movements so as to prevent them from occurring.

Some forms of commercial galvanometers are so arranged as to measure directly the product of the current and the difference of potential. Such instruments give the watts in the circuit and are, therefore, called wattmeters.

A form of wattmeter consists essentially of a thick wire coil placed in series in the circuit whose elec-

tric power is to be measured and a thin wire coil
placed as a shunt around the circuit. These two
coils, instead of acting on a needle, act on each other,
and the amount of their deflection will be proportional to the number of watts in the circuit.

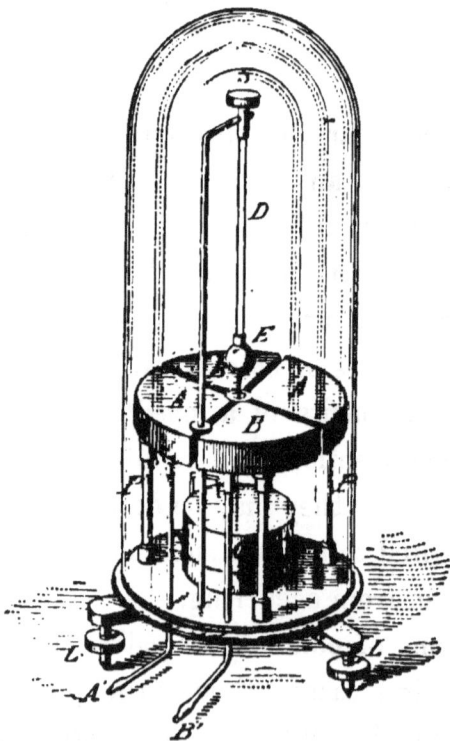

FIG. 19.—QUADRANT ELECTROMETER.

In the electrometer-voltmeter the difference of
potential is measured by electrostatic attractions and
repulsions. A well-known form of such instrument

is seen in the quadrant electrometer. In the quadrant electrometer the differences of potential are measured by the attractive and repulsive forces exerted by four plates or quadrants on a light, charged, needle of aluminium suspended within them.

A form of quadrant electrometer is shown in Fig. 19. The sectors of the quadrant are made of brass or other conducting metal shaped so as to form a hollow box. When placed together the four

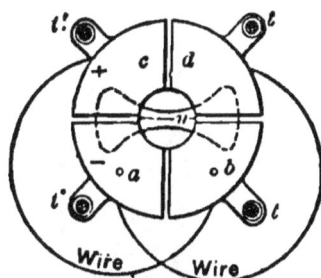

Fig. 20.—Quadrant Electrometer.

sectors are insulated from one another, but the opposite pairs are connected by a conducting wire, as shown in Fig. 20. A light aluminium needle *u*, is maintained at a constant electric charge by being connected with one of the coatings of a Leyden jar, or with one of the terminals of a sufficiently powerful voltaic battery. The electrometer needle is generally suspended by two parallel silk fibres so as to swing freely inside the hollow box quadrant. When at rest the needle has its greatest length exactly in the direction of the slot or space between the two opposite pairs of sectors, as shown in Fig. 20 by the dotted lines.

When, now, the two points of any circuit whose difference of potential is to be measured are con-

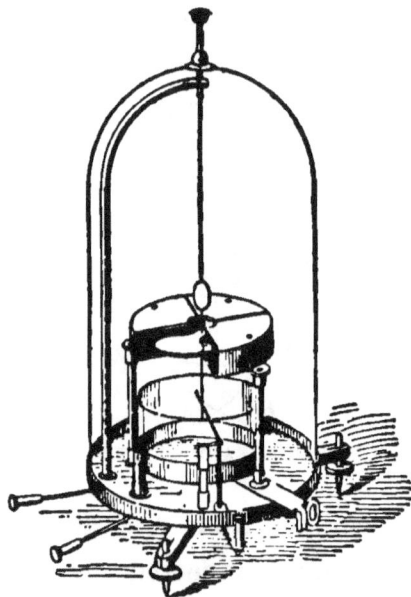

FIG. 21.—QUADRANT ELECTROMETER, SHOWING SUSPENDED NEEDLE.

nected to opposite pairs of quadrants, the charge so produced deflects the needle, and from the amount of this deflection the difference of potential between these points can be calculated. In the forms of electrometers shown in Figs. 19 and 21 the amount of this motion is determined by means of a spot of light reflected from a mirror E, supported by the suspension fibre, and generally observed through a telescope.

Electrometer-voltmeters are better adapted than galvanometer-voltmeters for determining differences of potential in certain cases because they do not require the passage of a current.

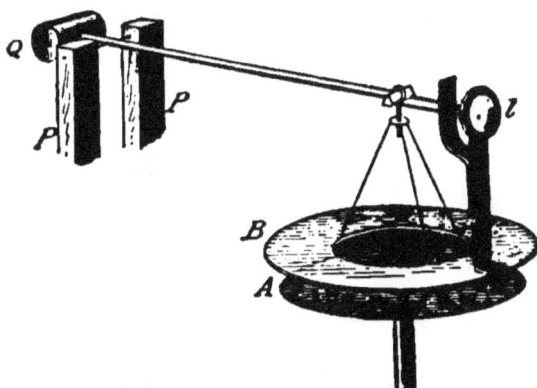

FIG. 22.—ATTRACTED DISC ELECTROMETER.

In the form of electrometer-voltmeter shown in Fig. 22, the difference of potential is measured by the weight required to balance the attraction which exists between two oppositely charged metallic discs. In the form here shown, the plate C, is suspended from the longer end of the lever l, within the fixed guard-plate or ring B, immediately above a second plate A, supported on an insulated stand. The plate

A, can be moved from or toward the plate *B*, through a measurable distance.

The electrostatic attraction is measured by the attraction of the fixed disc *A*, on the movable disc *C*. These two bodies are connected respectively to the two points whose difference of potential is to be determined. One of these may be the earth.

Instead of measuring the difference of potential directly by balancing the tendency to motion pro-

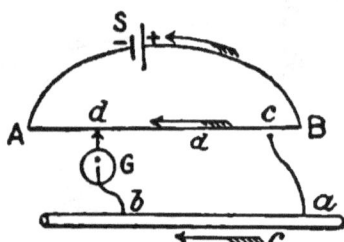

FIG. 23.—POTENTIOMETER.

duced by the attraction or repulsion by means of a weight, such differences may be determined by balancing or opposing an unknown difference of potential by a known difference of potential.

The apparatus shown in Fig. 23, called the potentiometer, is an apparatus of this character. The unknown difference of potential to be measured is balanced or opposed by a known difference of potential, the equality of the balancing being determined by the failure of one or more galvanometers, placed

In the shunt circuits, to show any movements of their needles.

A secondary battery S, or a standard voltaic cell, has its terminals connected to the ends A and B, of a wire of uniform diameter and of high resistance called the potentiometer wire. There will, therefore, occur along the wire a fall or drop of potential which will be equal per unit of length. This drop can be shown by connecting the terminals of a delicate galvanometer, generally of high resistance, to the different parts of the wire. The deflection of the needle will, of course, be greater the greater the length of wire between the two points touched.

If, now, the terminals of a standard voltaic cell, whose difference of potential is known, be connected so as to oppose the current taken from the potentiometer wire, and the contacts be slid along the potentiometer wire until no deflection of the needle is observed, the drop of potential between these points on the potentiometer wire will be equal to the difference of potential of the standard cell. In this way the wire is calibrated.

Suppose, now, it is desired to measure the difference of potential between two points a and b, on the wire C, through which a current is passing in the direction of the arrow. Connect the points b

and d, with a galvanometer G, of high resistance, and the points a and c, with a conductor; now slide c, toward or from d, until the galvanometer shows no deflection. The difference of potential between a and b is then equal to that between c and d.

The form of electrometer shown in Fig. 24 is called a capillary electrometer. In this instrument a horizontal glass tube, which is filled with mercury and has a drop of dilute sulphuric acid at B, has its ends connected with two vessels M and N, also

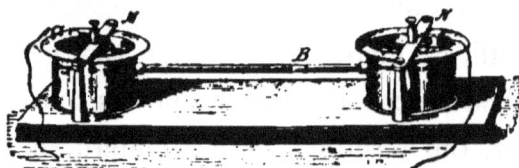

FIG. 24.—CAPILLARY ELECTROMETER.

filled with mercury. If the points whose difference of potential are to be determined are connected by means of conductors with the vessels M and N, a current passes through the capillary tube and a movement of the drop of acid takes place toward the negative pole. Provided the difference of potential does not exceed two volts, the amount of this movement is directly proportional to the difference of potential.

A ready means for obtaining a known difference of

potential, either for measuring an unknown difference of potential by opposing such known difference
of potential, or for the purpose of calibrating an instrument, and thus determining the value of its deflections, is found in various standard voltaic cells.
Ordinarily constructed voltaic cells would be impracticable for such purposes. With standard cells,
if especial care is taken to avoid polarization, and

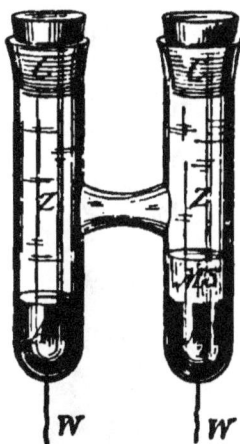

FIG. 25.—CLARK'S STANDARD CELL.

the circuit of the cell is closed but for a short time,
the electromotive force it produces is practically
constant. Standard voltaic cells are made in a great
variety of forms.

In Fig. 25 is shown a well-known form of standard voltaic cell devised by Latimer Clark, known as
the **H**-form of cell, from the shape of a vessel *C C*

connected by the cross tube. The left-hand vessel contains at A, an amalgam of pure zinc; the right-hand vessel contains at M, a mass of mercury covered with pure mercurous sulphate. Both vessels are then filled up to the level of the cross tube with a saturated solution of zinc sulphate to which a few

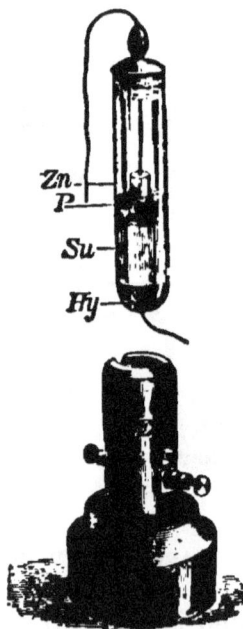

FIG. 26.—RAYLEIGH'S FORM OF CLARK'S STANDARD CELL.

crystals of the salt are generally added. Tightly fitting corks prevent loss by evaporation. Wires W, W, fused into the bottom of the vessels, serve as the terminals of the cell.

In Fig. 26 is shown Rayleigh's form of Clark's standard cell. The electrodes pass respectively through the bottom and top of a glass test-tube. On the bottom of the cell is placed a layer of mercury on top of which is placed a layer of mercurous sulphate paste that is rendered sufficiently semi-fluid by mixture with zinc sulphate to assume an approximately level surface.

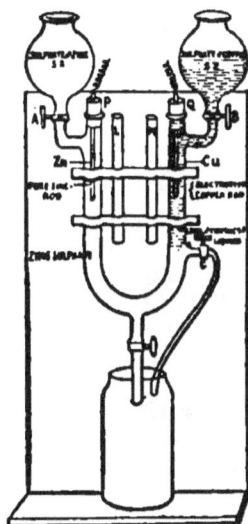

FIG. 27.—FLEMING'S STANDARD CELL.

The zinc is connected to the upper electrode and is immersed in this semi-fluid paste.

Fig. 27 shows Fleming's standard voltaic cell. The U-tube is connected by means of taps with two vessels filled respectively with chemically pure solutions of copper sulphate of the specific gravity of 1.1

at 15° C., and zinc sulphate of the specific gravity of 1.4 at 15° C.

To use the cell, the zinc rod *Zn*, connected with the wire passing through the rubber stopper, is placed in the left-hand branch. The tap at *A*, is opened and the entire U-tube is filled with the denser zinc sulphate solution. The tap at *C*, is then opened, and the liquid in the right-hand branch, above the tap, dis-

FIG. 28.—LODGE'S FORM OF DANIELL'S CELL.

charged into the lower vessel, but from this point only.

The tap *C*, is then closed and *B*, is opened, and the lighter copper sulphate is allowed to fill the right-hand branch above the tap *C*. The copper rod *Cu*, fitted to a rubber stopper and connected to a conducting wire, is then placed in the copper solution. Tubes at *L* and *M*, are provided for the reception of the zinc and copper rods when not in use. The copper rod is prepared for use by freshly electroplating it with copper. The electromotive force of this cell, when in proper action, is 1.074 volts. If the line of demarkation between the two liquids is not sharp, the arms of the vessel are emptied, and fresh liquid is run in.

Lodge's form of Daniell's standard cell is shown in Fig. 28.

Through a tube *T*, in a wide-mouthed bottle, is passed the glass tube, in the mouth of which is placed a zinc rod. A small test tube *t*, containing crystals of copper sulphate, is fastened to the bottom of the tube *T*, by means of a string or rubber band. The uncovered end of a gutta-percha insulated wire projects at the bottom *t*, through a tube in a tightly fitting cork, and forms the copper electrode. The bottle is filled with a solution of zinc sulphate.

The internal resistance of this form of standard cell is so high that it is only employed for use with measurements employing zero methods or with a condenser.

In any form of standard voltaic cell great care must be taken or otherwise appreciable differences in the electromotive force will result. When sulphate of copper is used care must be taken to prevent the copper from being deposited on the zinc. The temperature should be kept as nearly constant as possible. The electrolytes should be pure and kept at the same density. Where saturated solutions are employed, as in the case of the zinc sulphate employed in Clarke's **H**-form of cell, care should be taken to prevent the solution from becoming super-saturated.

A solution is saturated with a soluble salt when it contains as much as it can hold. When a saturated solution is cooled it deposits some of the salt, the liquid still remaining saturated. If, however, the liquid be closed from the air and cooled without shaking, the deposit of crystals may not occur, and the solution then becomes super-saturated.

EXTRACTS FROM STANDARD WORKS.

Gray, in his work entitled "The Theory and Practice of Absolute Measurements in Electricity and Magnetism,"* in describing electrometers, says on page 252 :

An electrometer has been defined as an instrument for measuring differences of electric potential by means of the effects of electrostatic force. It consists essentially of two conductors, between which is established the difference of potential which it is desired to measure. The electrostatic force set up produces motion of the parts of one of these conductors relatively to one another, or motion of the conductor as a whole relatively to the other conductor ; and from this motion, or from the mechanical force which must be applied to restore and maintain equilibrium in the configuration of zero electrification, the difference of potentials is inferred. We shall call this conductor the *Indicator* of the instrument.

When the instrument contains within itself a means of comparing the electric force called into play with other forces known in amount, as, for example, the force of gravity on a given mass, or the elastic reaction of a stretched spring, it gives directly by its indications the

* "The Theory and Practice of Absolute Measurements in Electricity and Magnetism," by Andrew Gray, M.A., F.R.S. London: Macmillan & Co. 1888. 518 pages, 105 illustrations. Price $3.75.

value in absolute electrostatic units of the difference of potential measured, and is called an *absolute electrometer.*

When the instrument gives only comparisons of the electrostatic forces with other forces, the amount of which it does not itself contain any means of determining, its indications can only be obtained in absolute units by a comparison with those of an absolute electrometer.

When the mode of variation of these undetermined forces is known and remains constant, one such accurate comparison is sufficient to allow a coefficient to be determined by which results proportional to measured difference of potential must be multiplied for reduction to absolute measure. The coefficient is called the *constant* of the instrument.

In a volume on " Primary Batteries," * on page 90, Carhart thus refers to the objections that exist as regards Lord Rayleigh's form of Clark's Standard cell :

The objections to Lord Rayleigh's form of the Clark normal element are: (1) the temperature coefficient is high and apparently variable : (2) it is not constructed in such manner as to keep the zinc and metallic mercury out of contact ; (3) the contact of the zinc and mercurial salt permits of local action whereby zinc replaces mercury.

Respecting the first objection, the method to be pursued in reducing the temperature coefficient is suggested by the

---

* "Primary Batteries," by Henry S. Carhart, A.M. Boston: Allyn & Bacon. 1891. 193 pages, 67 illustrations. Price $1.50.

fact, now well known, that the E. M. F. decreases with an increase in the density of the zinc sulphate solution. Hence, if the solution is saturated at 30° or 40°, upon a lowering of temperature the excess crystallizes out with a decrease of density. The reverse process takes place with rise of temperature, with the additional disadvantage that time is required for the diffusion of the redissolved salt. The temperature coefficient in such a cell is therefore made up of two parts ; one a real temperature effect, the other a secondary change resulting from a variability in the density of the zinc sulphate solution. A rise of temperature lowers the E. M. F. by increasing the density of the solution in addition to the direct primary effect of the temperature change.

The slowness of diffusion when the temperature rises makes the coefficient for a rapid change of temperature smaller than for a slow one. Thus Prof. Threlfall, investigating Clark cells made in accordance with Lord Rayleigh's directions, found the coefficient to be .000402 for a rapid rise of temperature from 21° to 34° C. This is less t`an half the value found by Lord Rayleigh between the same temperatures.

The magnitude of the temperature coefficient depends, then, upon the temperature at which the zinc salt is saturated, and, because of diffusion, upon the rapidity of the temperature change. To obviate these difficulties the zinc sulphate should be saturated at some definite temperature lower than any at which the cell is to be used. The temperature selected by the writer is that of melting ice. * * *

The other two objections urged against the usual form of

Clark cell are founded chiefly on the local action taking place when the zinc and mercurial salt are in contact. Zinc replaces mercury to some extent when in contact with a salt of mercury. With the oxide of mercury this action is very marked, resulting in the reduction of the mercury and oxidation of zinc. The same replacement process goes on with mercurous sulphate, zinc sulphate being formed at the expense of zinc and mercury sulphate, while the zinc is amalgamated with the reduced mercury. A progressive change in the density of the solution ensues, entailing perhaps a rise in the value of the temperature coefficient.

III.—THE MEASUREMENT OF ELECTRIC RESISTANCES.

In accordance with Ohm's law, when the difference of potential and the resistance of any circuit are known, the current strength in such circuit, or the number of coulombs per second, can be readily calculated. It is for this, and for many other reasons, a matter of great importance to be able to readily determine the resistance of any circuit, or part of a circuit.

Various methods can be employed for measuring electric resistances ; among the most important of these are the following:

(1.) By the method of substitution.

(2.) By a comparison of the deflections of a galvanometer.

(3.) By means of differential galvanometers.

(4.) By means of a resistance bridge in connection with a box of resistance coils.

(5.) By the indirect method of measuring the current and the electromotive force, and then calculating the resistance from the formula $R = \dfrac{E}{C}$.

(61)

In the method of substitution, the resistance to be measured x, Fig. 29, a box of resistance coils R and a galvanometer G, are placed in series in a circuit with a voltaic battery B, by means of conductors that are sufficiently thick to permit their resistance to be neglected. The deflection of the galvanometer is first obtained with the known resistance x, only in the circuit ; that is, with no resistance in the box B, which is effected by inserting all its plug keys.

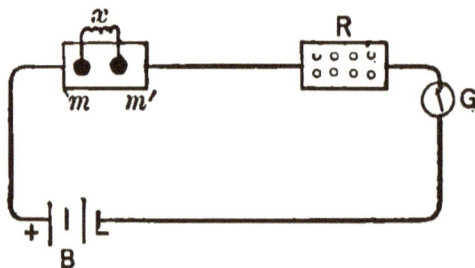

FIG. 29.—SUBSTITUTION METHOD.

The resistance x, is then cut out of the circuit by placing a thick copper wire $m\,m'$, across x, so as to short-circuit it. Resistances are then unplugged in the box R, until the same deflection is obtained in the galvanometer, when, provided the difference of potential of the battery has remained constant, the resistance unplugged will equal the unknown resistance.

The resistance box employed in the above method

consists generally of a number of coils of wire, the electric resistance of which is accurately known. In order to prevent the magnetic field produced by such coils, when traversed by an electric current, from affecting the needle of a galvanometer placed near them, they are wound after the wire which forms them has been doubled or bent on itself in two equal lengths so that the two halves extend parallel with each other.

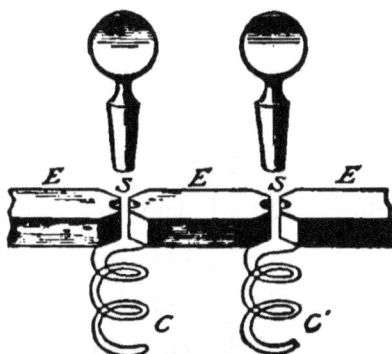

Fig. 30.—Resistance Coils.

The arrangement of the coils which form the resistance, as well as this method of winding, are shown in Fig. 30. The coils C C', are connected to each other by means of thick pieces of brass E, E, E, to which their ends are soldered. When the plug keys are placed in the holes or spaces left at S, S, between the contiguous brass pieces E, E, E, the coils are cut out of the circuit by short-circuiting. When,

however, the keys are removed, the coils are placed
in the circuit by what is technically called unplugging.

In order to avoid changes in the electric resistance
of the coils, on changes in temperature, the coils are
generally made of German silver wire, or, preferably,
of platinoid, or an alloy of platinum and silver, the
resistance of which is not sensibly affected like that
of most other metals by changes in temperature.
Even, however, in such cases it is necessary not to
permit the current to flow through the coils for
longer than a few minutes at a time.

The method of determining the value of a resistance by a comparison of the deflection of galvanometers is based on the fact that such deflections
are proportional to the current passing, and, if the
electromotive force of the electric source employed
is constant, that the current which passes will decrease as the resistance increases.

The method of determining resistances by the use
of differential galvanometers is based on balancing
the deflection of the needle of a galvanometer placed
in the circuit of one of the coils in which the unknown resistance is placed, by the opposing magnetic
effects of the other coil in which a known resistance
is placed.

By far the preferable method of determining resistances is by means of a device invented by Wheatstone, called the electric bridge; or, as it was originally called, the electric balance, because in it a known resistance is balanced against an unknown resistance.

The operation of the electric bridge or balance is based on the fact that no current will flow through a conductor the terminals of which are connected to points that are of the same difference of potential.

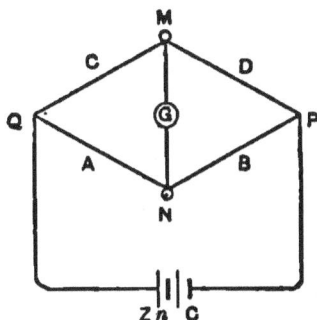

FIG. 31.—ELECTRIC BRIDGE.

A simple form of electric bridge is shown in Fig. 31. *A, B, C, D,* are electric resistances called the arms of the bridge. Any of these resistances can be determined, provided the absolute value of one and the relative values of the other two are known.

These resistances are arranged in the manner shown, and the terminals *Zn* and *C,* of the voltaic battery, are connected to the points *Q* and *P.* The current of the battery then flows through the circuit, branches at P, part flowing through the arms D and C, and the remainder flowing through the arms B and A. These two branches unite at Q, and return to the battery. A sensitive galvanometer, G, is placed in the circuit connecting the points M and N. The name bridge was derived from the fact that this wire or conductor bridges or connects the points M and N.

When a current passes through any conductor, a drop or fall of potential occurs that is proportional to the resistance. When, therefore, the current from the voltaic battery, $Zn\ C$, branches at P, and passes through the resistances D and C, and B and A, a drop or fall of potential occurs along the paths D and C, and B and A, proportional to their resistances. If the points M and N, that are connected by the bridge wire, are at the same difference of potential, no current will pass ; if, however, they are at a different potential, a current will pass from M, to N, if M, is at the higher potential, and in the opposite direction, from N, to M, if M, is at the lower potential, and the needle of the galvanometer will be deflected according to the direction in which the current flows. If, therefore, the known resistances A, C and B, are so proportioned to the value of the unknown resistance D, that no current passes through the galvanometer G, between the points M and N, then such points are at the same difference of potential, and since the fall of potential is proportional to the resistance, it follows that

$$A : B :. C : D,$$

$$A \times D = B \times C,$$

$$\text{or } D = \left(\frac{B}{A}\right) C.$$

If, then, the values A, B and C, are known, the value of D, can be readily calculated.

By making the value $\dfrac{B}{A}$ some simple ratio, the value of D, is easily obtained in terms of C.

The resistances A, B and C, may consist of coils of wire whose resistances are unknown.

There are two forms of Wheatstone bridge ; namely, the box form and the sliding form. In the box form, the arms of known resistance of the bridge consist of resistance coils. In the sliding form one of the known resistances consists of a resistance coil, and the other two of a uniform wire or conductor over which a sliding contact moves so as to place different lengths of the conductor on different sides of the slide, and, therefore, in different arms of the bridge.

The box form of bridge is shown in perspective in Fig. 32 and in plan in Fig. 33.

It will be noticed that the bridge arm corresponding to the resistances A and B, of Fig. 31, consists

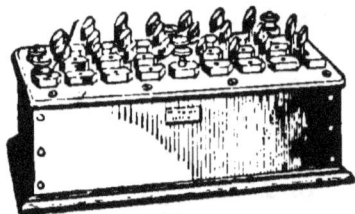

FIG. 32.—BOX BRIDGE.

of resistance coils of 10, 100, 1,000 ohms each, called the proportionate coils. The arm corresponding to the resistance C, of the same figure, is composed of separate resistances 1, 2, 2, 5, 10, 10, 20, 50, 100, 100, 200, 500, 1,000, 1,000, 2,000, and 5,000 ohms

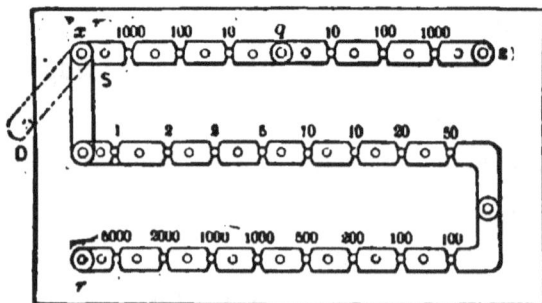

FIG. 33.—BOX BRIDGE.

The following are the connections: the galvanometer is inserted between q and r, Fig. 34; the unknown resistance between the z and r, and the battery is connected to x and z. A definite proportion being taken for the value of the proportionate coils, such, for example, as 10 on one side, and 100 on the other, resistances are inserted in the arm D, until the galvanometer G, shows no deflection.

The similarity between these connections and those shown in Fig. 31, will be recognized from an inspection of Fig. 34. The arms A and B, of Fig. 31, correspond to the arms $q\,x$ and $q\,z$, of Fig. 34 ; the arm C, of Fig. 31, corresponds to the arm $x\,r$,

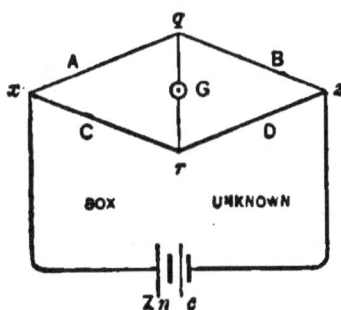

FIG. 34.—ELECTRIC BALANCE.

of Fig. 34, and D, to the unknown resistance. Then as before we have :

$$A : B :: C : D, \text{ or } A \times D = B \times C. \therefore D = \left(\frac{B}{A}\right) C.$$

The advantage of the simplicity of the ratios A and B, or 10, 100 and 1,000 of the bridge box, will therefore be manifest. The battery terminals may be connected to q and r, and the galvanometer terminals to x and z, without disturbing the above proportions.

In the slide form of bridge, as shown in Fig. 36, the proportionate arms are formed of a single thin wire of high resistance and uniform diameter, formed of German silver or platinoid.

The slide contact key moves over the wire, one terminal of the key being connected with the galvanometer and the other, when the key is depressed, with the wire. From the uniform diameter of the

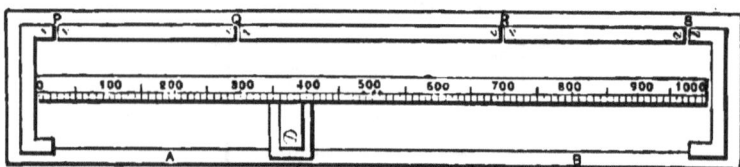

FIG. 35.—SLIDE BRIDGE.

wire, the resistance on either side of the key will be directly proportional to the length of the wire. A graduated scale placed under the wire serves to measure its length. A thick metal strip connected with the slide wire has four gaps at P, Q, R and S. When in ordinary use the gaps at P and S, are either connected by stout strips of conducting material, or by known resistances; in the latter case they act as ungraduated extensions of the slide wire, and,

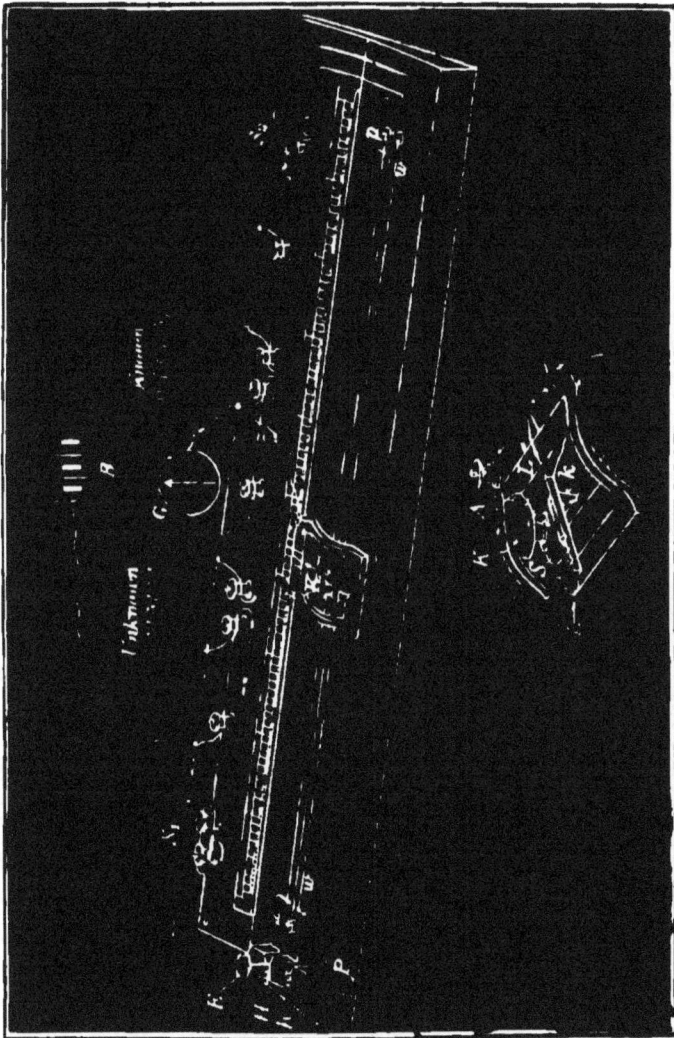

FIG. 36.—SLIDE BRIDGE.

like lengthening the slide wire, increase the sensibility of the bridge.

The unknown resistance is then inserted in the gap at Q, and the known resistance, generally a resistance box, in the gap at R. The galvanometer has one of its terminals connected to the metal strips between Q and R, and its other terminal to the sliding key. The battery terminals are connected to the metal strips between P and Q, and R and S, respectively.

These connections are more clearly seen in the form of bridge shown in Fig. 36. The slide wire w w, consists of three separate wires each a meter in length, and so arranged that either only one wire or two in series can be used. The circuit being now arranged as shown, the sliding key is moved until no current flows through the galvanometer when the key is depressed.

The slide form of bridge is not entirely satisfactory, since the uncertainty of the spring contact causes a lack of correspondence between the contact and the points of the scale on which the index rests.

The loss of uniformity in the diameter of the wire, due to constant use, also causes a lack of correspondence between the resistance of the wire and its length. With care, however, very accurate results can be obtained by the slide bridge.

For the purpose of standardizing resistance coils, a standard coil of some known value is necessary. A form of standard ohm, issued by the Committee of Electrical Standards in England, is shown in Fig. 37. A coil of platinum silver wire, insulated by silk covering and melted paraffine, is placed at *B*, and the space above it at *A*, is filled with paraffine except at *t*, where an opening is left for the insertion

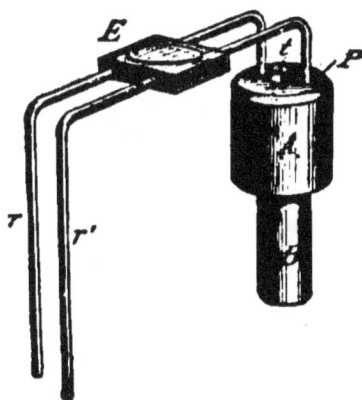

FIG. 37.—STANDARD OHM.

of a thermometer. The ends of the resistance coil are soldered to thick copper rods *r r'*, as shown.

In the resistance box already described the adjustability of the resistance is obtained by means of unplugging coils of various values until the required resistance is introduced into the circuit. Wheatstone devised an apparatus in the shape of an adjustable resistance or rheostat of the form shown in Fig. 38.

This instrument is suitable for rough work, but is not very satisfactory where great accuracy is required. It consists, as shown, of two parallel cylinders *A* and *B*, formed respectively of conducting and non-conducting material. The resistance wire, which is a bare wire, can be wound from the surface of the non-conducting cylinder *B*, to the conducting cylinder *A*. Only the wire on *B*, is introduced into

FIG. 38.—WHEATSTONE'S RHEOSTAT.

the circuit, since the bare wire on the conducting cylinder *A*, is short circuited by the metallic cylinder.

As a ready means for measuring the resistance of a circuit through which a current of electricity is passing Ayrton has devised an instrument called an ohmmeter, which shows directly, by the deflection of a magnetic needle, the resistance in any part of a circuit.

The construction of the apparatus will be understood from an inspection of Fig. 39. Two coils of wire, $C C$ and $c c$, formed respectively of short thick wire and long thin wire, are placed at right angles to each other, and act on a soft iron needle placed as shown. The short thick coil $C C$, is connected in series to the conductor O, whose resistance is to be measured. The long thin coil $c c$, of known high resistance, is placed as a shunt to O.

Under these circumstances the action of the needle is due to the ratio of the difference of potential at

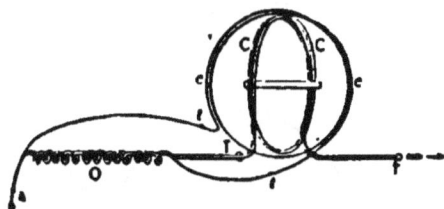

FIG. 39.—AYRTON'S OHMMETER.

the terminals of the unknown resistance and the current strength in the thick coil, or $R C = E$, as may be deduced from Ohm's law.

The coils are so proportioned that when the current flows through the short thick wire it moves the needle to the zero of the scale, while the long thin wire produces a deflection directly proportional to the resistance.

When a conductor is in the form of a bare wire of uniform area of cross section, the resistance of a

given length of such conductor can be readily calculated from a knowledge of its diameter, provided,
of course, the specific resistance of such material—
that is, the resistance of a unit length of unit area
of cross section—is known.

In order to readily determine the diameter of a
wire so as to calculate the resistance of a given
length of such wire, various instruments called wire
gauges are employed.

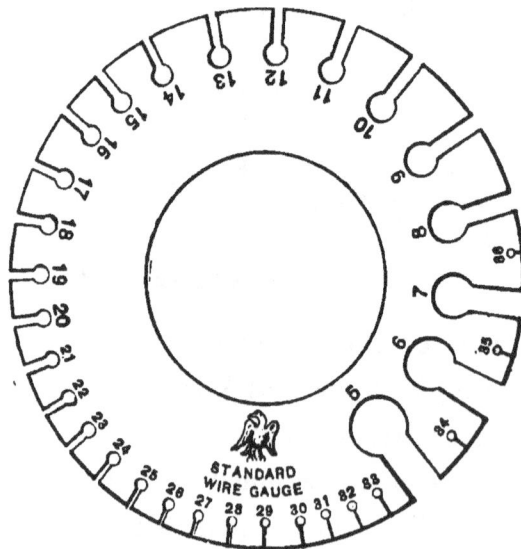

FIG. 40.—ROUND WIRE GAUGE.

Fig. 40 shows a form of wire gauge called the
round-wire gauge. In this gauge notches of varying
width, cut in the edges of a circular plate of steel,
serve to approximately measure the diameter of a

wire which is passed lengthwise through the openings. Numbers indicating the different sizes of the wire are affixed to each of the openings.

Another form of wire gauge, called the micrometer wire gauge, is shown in Fig. 41.

The wire to be measured is placed between the fixed support *B*, and the movable end *C*, of a long screw, which accurately fits a threaded tube *a*. A thimble *D*, provided with a milled head, fits over the

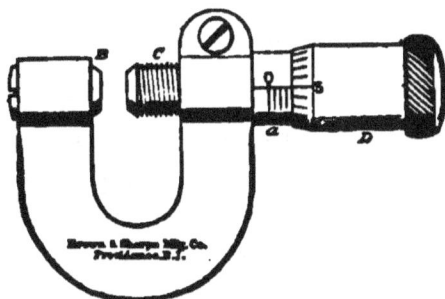

FIG. 41.—VERNIER WIRE GAUGE.

screw *C*, and is attached to the upper part. The lower circumference of *D* is divided into a scale of 20 equal parts. The tube *a* is marked in divisions equal to the pitch of the screw. Every fifth of these divisions is marked as a larger division.

The principle of operation of the gauge is as follows: Suppose the screw has 50 threads to the inch; the pitch of the screw, or the distance between two contiguous threads, is, therefore, $\frac{1}{50}$, or .02, of an inch.

One complete turn of the screw will, therefore, advance the sleeve D, over the scale a, the .02 of an inch. If the screw is only moved through one of the twenty parts marked on the end of the thimble or sleeve parts, or the $\frac{1}{20}$ of a complete turn, the end C, advances toward B, the $\frac{1}{20}$ of $\frac{1}{50}$, that is, $\frac{1}{1000}$, or .001, inch.

Suppose, now, a wire is placed between B and C, and the screw advanced until the wire fairly fills the space between B and C, and the reading shows two of the larger divisions of the scale a, three of the smaller ones and three on the end of the sleeve D, then

Two large divisions of scale a = .2 inch.

Three smaller divisions of scale a = .06 inch.

Three divisions on circular scale on D = .003 inch.

Diameter of wire = .263 inch.

In the self registering wire gauge the apparatus is arranged to give directly without calculation the exact diameter of the wire to be measured.

A form of self-registering wire gauge is shown in Fig. 42. The wire or plate is inserted in the gap between a fixed and a movable plate. The diameters of the wire or plate that is being measured are then read off, shown on one side of the gauge, and the gauge numbers on the other.

In making calculations of the resistance of any circuit it must carefully be remembered that changes in temperature produce changes in resistance. The character of such changes can be summarized as follows:

FIG. 42.—WIRE AND PLATE GAUGE.

(1.) The electric resistance of metallic conductors increases as the temperature rises. The resistance of the carbon conductor of an incandescent lamp decreases as the temperature rises. Its resistance decreases when it is raised to incandescence, such decrease amounting to about three-eighths of its resistance when cold.

Roughly speaking the increase in temperature is proportional to the temperature. In reality, the resistance of all metals except mercury increases more rapidly than the temperature.

CHEMICALLY PURE SUBSTANCES ARRANGED IN ORDER OF INCREAS-
ING RESISTANCE FOR THE SAME LENGTH AND SECTIONAL AREA.
LEGAL MICROHM.

NAME OF METAL.	Resistance in microhms at 0° C.		Relative resist-ances.
	Cubic centi-metre.	Cubic inch.	
Silver, annealed	1.504	0.5921	1.
Copper, annealed	1.598	0.6292	1.063
Silver, hard drawn....	1.634	0.6433	1.086
Copper, hard drawn..	1.634	0.6433	1.086
Gold, annealed............................	2.058	0.8102	1.369
Gold, hard drawn....	2.094	0.8247	1.393
Aluminium, annealed........	2.912	1.147	1.935
Zinc, pressed	5.626	2.215	3.741
Platinum, annealed......................	9.057	3.565	6.022
Iron, annealed	9.716	3.825	6.460
Gold-silver alloy (2 oz. gold, 1 oz. silver), hard or annealed..........................	10.87	4.281	7.228
Nickel, annealed.......	12.47	4.907	8.285
Tin, pressed.......	13.21	5.202	8.784
Lead, pressed............................	19.63	7.728	13.05
German silver, hard or annealed..........	20.93	8.240	13.92
Platinum-silver alloy (1 oz. platinum, 2 oz. silver), hard or annealed....	24.39	9.603	16.21
Antimony, pressed......................	35.50	13.98	23.60
Mercury................................	94.32	37.15	62.73
Bismuth, pressed....	131.2	51.65	87.23

Dr. Matthiessen's investigations as to the resistance of various metals, and the effects of temperature thereon, give the resistance of certain cubic

volumes, such for example, as one cubic centimetre, or one cubic inch of such metals in michroms.

In the preceding table, from Ayrton, calculated
from Matthiessen's results, the resistance in microhms
is given of a cubic centimetre and a cubic inch respectively of the various metals. In the last column
will be found the relative resistance as compared
with silver. The resistance is measured laterally
across the cube from one face to the opposite face.

(2.) The resistance of an electrolyte decreases as
temperature increases.

(3.) The resistance of di-electrics or non-conductors decreases as the temperature increases.

The resistance of a wire is directly proportional
to its length, inversely proportional to the area of
cross-section, or to the square of the diameter, and
depends on the material of which the wire is formed ;
calling R, the resistance, l, the length, d, the diameter, and K, a constant, varying with the material,
then $R = \left(\dfrac{2}{d^2}\right) K$.

In the following table from Jenkin the conducting power, or resistance in ohms, is given for a number of metals. The conducting power is compared
with that of pure hard drawn silver wire, which is
taken as one hundred.

TABLE OF CONDUCTING POWERS AND RESISTANCES IN OHMS.

NAMES OF METALS	Conducting power at 0 degree C.	Resistance of a wire one foot long weighing one grain.	Resistance of a wire one metre long weighing one gramme.	Resistance of a wire one foot long $\frac{1}{1000}$ inch in diameter.	Resistance of a wire one metre long, one milli-metre in diameter.	Approximate percentage of variation in resistance for 1 degree of temperature at 20 deg. C.
Silver, annealed....	......	0.2214	0.1544	9.936	0.(1937	0 377
Silver, hard drawn	100.00	0.2421	0.1689	9.151	0.02103	......
Copper, annealed..	....	0.2064	0.144	9.718	0.02057	0.388
Copper, hard drawn	99.55	0.2105	0.1469	9.94	0.02104	......
Gold, annealed......	....	0.5849	0.408	12.52	0.0265	0.355
Gold, hard drawn..	77.96	0.295	0.415	12.74	0.02697	......
Aluminium, annealed	....	0.06822	0.05759	17.72	0.03751	......
Zinc, pressed	29.02	0.571	0.3983	32.22	0.07244	0.365
Platinum, annealed	....	3.526	2.464	55.09	0.1166	......
Iron, annealed	16.81	1.2425	0.7522	59.40	0.1251	......
Nickel, annealed..	13.11	1.0785	0.8666	75.78	0.1604	......
Tin, pressed	12.36	1.317	0.9184	80.36	0.1701	0.365
Lead, pressed	8.32	3 236	2.257	119.39	0.2527	0.387
Antimony, pressed.	4.62	3.321	2.3295	216.0	0.4571	0.389
Bismuth, pressed...	1.24	5.054	3.575	798.0	1 689	0.354
Mercury, liquid....	....	18.74	13.071	600.0	1.27	0.072
Platinum-silver, alloy, hard or annealed	......	4.243	2.959	143.35	0.3140	0.031
German silver, hard or annealed	....	2.652	1.85	127.32	0.2695	0.044
Gold, silver alloy, hard or annealed..	....	2.391	1.668	66.1	0.1399	0.065

—*Jenkin.*

EXTRACTS FROM STANDARD WORKS.

Kempe in his " Handbook of Electrical Testing,"* on page 10, speaks of resistance coils as follows :

The essential points of a good set of resistance coils are, that they should not vary their resistance appreciably through change of temperature, and that they should be accurately adjusted to the standard units, which adjustment ought to be such that not only should each individual coil test according to its marked value, but the total value of all the coils together should be equal to the numerical sum of their marked values. It will be frequently found in imperfectly adjusted coils that although each individual coil may test, as far as can be seen, correctly, yet when tested altogether their total value will be one or two units more or less than the sum of their individual values ; because, although an error of a fraction of a unit may not be perceptible in testing each coil individually, yet the accumulated error may be comparatively large.

The wire of the coils is, as a rule, of German silver, the specific resistance of which metal is but little affected by variations of temperature. The wire is usually insulated by two coverings of silk and is wound double on ebonite bobbins, the object of the double winding being to eliminate the extra current which would be induced in the coils if the wire were wound on single. By double winding the

---

* "A Hand Book of Electrical Testing," by H. R. Kempe. Fifth edition. London: E. & N. Spon. 1892. 576 pages, 200 Illustrations. Price, $7.25.

current flows in two opposite directions on the bobbin, the
portion in one direction eliminating the inductive effect of
the portion in the other direction. When wound, the bob
bins are saturated in hot paraffine wax, which thoroughly
preserves their insulation, and prevents the silk covering
from becoming damp, which would have the effect of partially short-circuiting the coils and thereby reducing their
resistance.

The small resistances are made of thick wire, the higher
ones of thin wire to economise space

When bulk and weight are of no consequence, it is better
to have all the coils made of thick wire, more especially if
high battery power is used in testing, as there is less
liability of the coils to become heated by the passage of
the current through them.

IV.—VOLTAIC CELLS.

When Luigi Galvani, in 1786, first announced his classic experiment with the frog's legs, it was generally believed that he had discovered the cause of vitality. Alexander Volta, like most of the scientific men of his time, adopted this view, until a more careful study led him to see that what actually caused the convulsive movements of the frog's legs was electricity, and that Galvani had discovered, not the cause of vitality, but a new method of producing electricity.

Volta, repeating the experiments of Galvani, found that the movements of the frog's legs were more pronounced when the muscles were brought into contact with the nerves by means of two dissimilar metals, such as zinc and copper connected at one pair of ends and brought into contact with the nerves and muscles at the other pair of ends, as shown in Fig. 43.

As the result of extended experiments in 1800, Volta conceived of an apparatus for the production of electricity, which, in honor of its inventor, was named the voltaic pile.

Volta ascribed the origin of the electricity in the pile, as well as in the experiment with the frog's legs, to the contact of dissimilar metals; and, although this view is still held by some, it is now generally believed that while the mere contact of

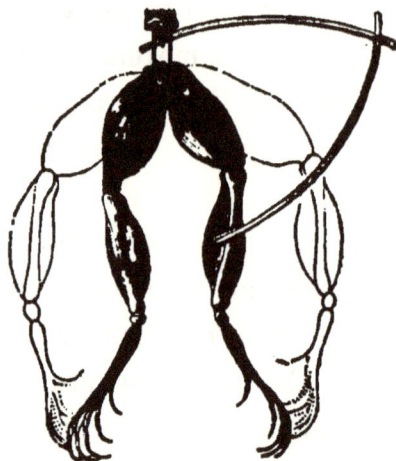

FIG 43.—GALVANOSCOPIC FROG.

dissimilar metals will produce differences of potential, it is to the gradual oxidization of one of the metals, in this case the zinc, that that energy is liberated which maintains an electric current as long as any chemical action continues.

A form of voltaic pile, similar to Volta's original pile, is shown in Fig. 44. It consists of alternate

discs of copper, zinc and wet cloth placed in a vertical pile one over the other. The top and bottom of
the pile is formed of a plate of copper and a plate of

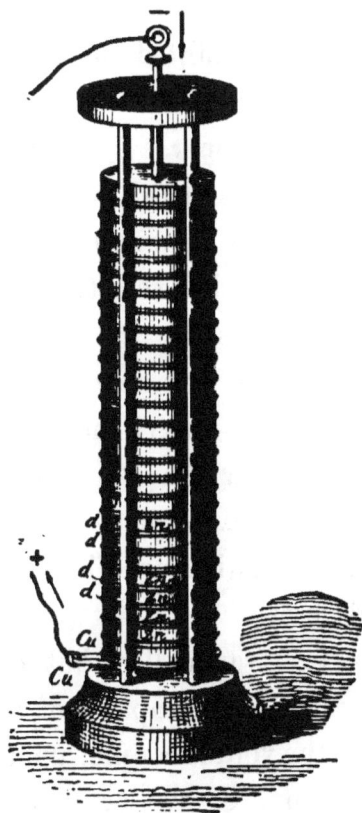

FIG. 44.—VOLTAIC PILE.

zinc respectively, which form the poles of the battery.

Any combination of parts by means of which electricity can be produced in this manner by chemical

action is called a voltaic cell. The voltaic cell is by some called the galvanic cell. When, however, it is remembered that Galvani originally claimed the discovery of a vital fluid, or principle of life, and that it was Volta who first pointed out the true cause of the electricity produced, it will be seen that the term galvanic cell is a misnomer.

A voltaic cell consists of two parts; namely,

(1.) Of a voltaic pair or couple.

(2.) Of a liquid called the electrolyte, in which the voltaic couple is immersed.

The voltaic couple or pair generally consists of two dissimilar metals. It may, however, consist of couples or pairs formed of a great variety of different substances; such, for example, as metals and metalloids, different gases, different liquids, or of different liquids and gases.

The electrolyte is generally, but not always, an acid solution. It must be capable of acting on one of the metals of the voltaic couple, and of conducting and of being decomposed by electricity.

If a plate of zinc and a plate of copper be placed in a dilute solution of sulphuric acid and water, and left unconnected outside of the liquid, then the following phenomena take place :

(1.) The sulphuric acid is decomposed, a salt of

zinc called zinc sulphate is formed, and hydrogen gas is liberated.

(2.) The hydrogen is liberated mainly at the surface of the zinc plate.

(3.) The entire mass of the liquid becomes heated.

If, however, the plates are connected outside of the battery by a conductor, then the sulphuric acid is decomposed as before, but the remaining phenomena are changed ; for,

(1.) The hydrogen is now liberated mainly at the surface of the copper plate.

(2.) The heat does not appear in the liquid only, but is distributed throughout all parts of the circuit.

(3.) An electric current is now produced which flows through the entire circuit, and will continue to so flow as long as any sulphuric acid remains to be decomposed, and zinc to unite with the acid ; that is, as long as any chemical action takes place.

Very clearly, then, the energy which formerly appeared as heat in the liquid now appears in all parts of the electric circuit as electric energy.

We may, therefore, regard the true cause of the production of electricity in the voltaic cell as due to the combination of the zinc with the sulphuric acid, and not to the contact of two dissimilar metals.

In any voltaic couple consisting of two dissimilar metals, one of the metals is acted on by the electrolyte while the other is left untouched. In the case of the zinc-copper couple immersed in sulphuric acid, the metal which is acted on by the liquid will be the zinc.

When a zinc-copper voltaic couple is immersed in sulphuric acid, and the circuit is completed outside the battery by a conductor, the electric current pro-

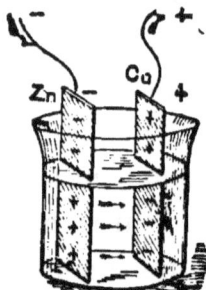

FIG. 45.—VOLTAIC COUPLE.

duced will flow through the liquid from the zinc plate to the copper plate, and through the circuit external to the liquid, from the copper plate to the zinc plate, re-entering the cell at the zinc plate.

In accordance with the convention already made, namely, that an electric current flows out of a source at its positive pole and re-enters it at its negative pole, it will be seen that, in such a couple as that shown in Fig. 45, the part of the copper plate which is outside the liquid will form the positive pole of the cell, and that the similar part of the zinc plate will form the negative pole. By the same reasoning, however, that portion of the zinc plate which is covered by the liquid will have a current of electricity flowing out of it, and will, therefore, form the positive pole, while the similar portion of the copper plate will form the negative pole.

In other words, the parts of the metallic plates that are covered by the electrolyte are to be regarded as of different polarity from the parts which project from the liquid.

In point of fact the zinc plate is all of the same polarity, namely, negative, as an electrometer attached to this plate would show. The apparent positive polarity of the plate under the liquid is probably due to the polarity of that part of the electrolyte which touches the zinc plate.

The ease with which different metals form voltaic couples with zinc renders it necessary to obtain plates of chemically pure zinc for use in voltaic cells, since otherwise minute voltaic couples will be formed by the particles of the impurities present, and the strength of the cell will be wasted in producing currents in minute closed circuits. As it is practically impossible to obtain plates of chemically pure zinc it is necessary to cover the surface of the zinc plates with an amalgam of zinc and mercury. This process of covering the zinc plate is called the amalgamation of the plate, and is easily obtained by dipping the plate for a few moments in dilute sulphuric acid and then rubbing a small quantity of mercury over the surface. The mercury adheres to the surface of the zinc plate, covering it with a bright coating of amalgam.

During the action of the electrolyte on the positive plate of a voltaic couple the hydrogen tends to be liberated at the surface of the negative plate, and this plate will finally become coated with a covering of hydrogen, unless some means are taken to avoid it.

It is very important to understand the effect which this film of hydrogen produces on a voltaic cell. It will be remembered that the current generated in the voltaic cell flows from the positive plate through the liquid to the negative plate. But hydrogen is more positive than zinc, and tends, therefore, to produce an electric current in the opposite direction to that produced by the zinc ; namely, from the copper plate to the zinc plate.

The hydrogen gas does not actually produce this current ; it only tends to produce it ; or, in other

words, the hydrogen produces what is called a counter-electromotive force, and the cell undergoes what
is called polarization. The polarization of a voltaic
cell causes a weakening of the current produced, for
the following reasons :

(1.) On account of the counter-electromotive force
produced, which causes a decrease in the effective
electromotive force of the cell.

(2.) On account of the increased resistance of the
voltaic cell, due to the collection on the surface of
the plate of bubbles of gas, which possess a high resistance.

There are three ways in which the ill effects of
polarization may be avoided, namely :

(1.) Mechanically. The bubbles of gas are brushed
off the surface of the negative plate by means of a
stream of air or of liquid. Or, they are permitted
to pass off by roughening the surface of the plate
and thus covering it with points.

(2.) Chemically. The surface of the negative
plate is surrounded by some powerful oxidizing substance like nitric or chromic acid, which is capable
of oxidizing the hydrogen and thus removing it from
the plate.   .

(3.) Electro-chemically. The negative plate is immersed in a solution of the same metal as that of

which it is composed. For example, if copper forms the negative plate of the cell, it is immersed in a solution of copper sulphate, so that the hydrogen, which tends to be liberated at the surface of the copper plate, in passing through the solution of the copper salt surrounding this plate, decomposes the salt and deposits a film of copper over the plate.

Although there are a great variety of the voltaic

FIG. 46.—SMEE CELL.

cells, yet they may readily be divided into two great classes; namely,

(1.) Single-fluid cells.

(2.) Double-fluid cells.

In single-fluid cells, as the name implies, the voltaic couple or pair is dipped into a single electrolytic fluid, while in the double-fluid cell each element or plate of the couple is dipped into a different fluid.

As a rule, double-fluid cells are less liable to polarization and are capable of giving a constant current for a longer time than single-fluid cells.

The well-known form of single-fluid cell shown in Fig. 46 is called, after its inventor, the Smee cell. The voltaic couple or pair is formed of zinc and silver. The silver plate is generally placed between

FIG. 47.—BICHROMATE CELL.

two plates of zinc. The electrolyte employed is dilute sulphuric acid.

The silver plate has its surface roughened by a covering or coating of platinum in a finely divided state known as platinum black. This cell produces an electromotive force of about .65 volt.

Another well-known form of single-fluid cell, shown in Fig. 47, is known as the bichromate cell. The voltaic couple consists of zinc and carbon. The electrolyte is formed by dissolving one pound of bichromate of potash in ten pounds of water and gradually adding to the solution two and one-half

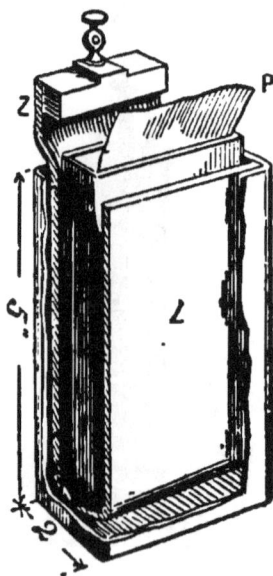

FIG. 48. — GROVE'S NITRIC ACID CELL.

pounds of ordinary commercial sulphuric acid. The electromotive force of this cell is about 1.9 volts. The cell is capable of giving a strong current; but, since it readily polarizes, it only furnishes a constant current for a comparatively short time.

In double-fluid cells, since there are two separate

liquids used as electrolytes, some means must be adopted for keeping these liquids separate. This is generally accomplished by the use of a jar of unglazed earthenware, called a porous jar or cell.

Grove's nitric acid cell, shown in Fig. 48, is a well-known form of double-fluid cell. The voltaic couple is formed by plates of zinc and platinum. The platinum is placed in a porous jar containing

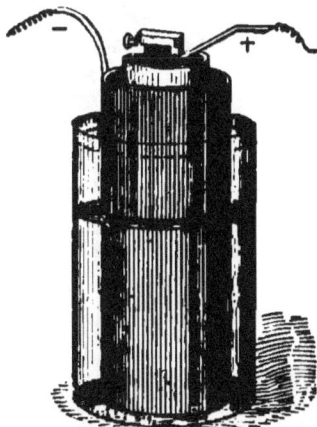

FIG. 49.—BUNSEN CELL

nitric acid, and the zinc in a cell containing dilute sulphuric acid. The nitrous fumes which this cell gives off during action render its use unpleasant. It gives an electromotive force of about 1.93 volts.

The Bunsen cell, shown in Fig. 49, is a modification of the Grove cell, in which the platinum is replaced by a plate of carbon. As in the Grove cell, the zinc is dipped into dilute sulphuric acid and the carbon in strong nitric acid. This cell gives an electromotive force of about 1.96 volts.

The great objection to all double-fluid cells, thus far described, is to be found in the fact that the current which they supply is far from constant. As the chemical action goes on, the strength of the acid

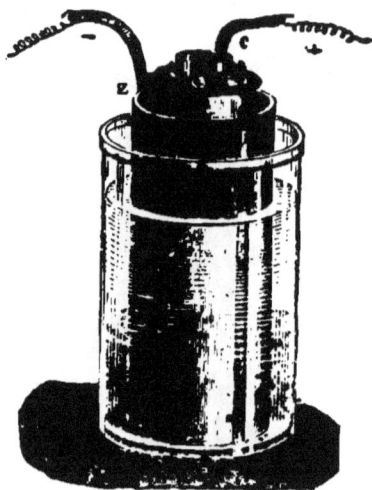

FIG. 50.—DANIELL'S CONSTANT CELL.

electrolytes greatly decreases, and a corresponding decrease occurs in the current strength. The problem of obtaining a constant electric current from a voltaic cell was solved by Prof. Daniell, who produced the cell named after him, and shown in Fig. 50.

In Daniell's constant cell the voltaic couple is formed of zinc and copper dipped respectively into dilute solutions of sulphuric acid and a saturated solution of copper sulphate or bluestone.

The sulphuric acid is placed in the outer cell and the copper sulphate inside the porous cell. A perforated cage or vessel, kept filled with crystals of bluestone, is supported so as to have the lower portions of the crystals continually in contact with the liquid.

The action on which the constancy of the Daniell cell depends is as follows : As the sulphuric acid of the electrolyte enters into combination with the zinc, the hydrogen, which is thereby set free, passes through the porous jar, but before it reaches the surface of the copper plate it meets the solution of copper sulphate surrounding this plate, and, decomposing it, deposits metallic copper on its surface and sets free or liberates sulphuric acid, which passes through the pores of the porous cup into the outer jar. As the strength of the solution of copper sulphate decreases, from its gradual decomposition, enough crystals of the salt are dissolved from the cage in the upper part of the liquid to keep the solution saturated. The invention by Daniell of this cell rendered telegraphy commercially possible.

The Daniell cell gives an electromotive force of about 1.072 volts.

A serious objection to the use of the Daniell cell is to be found in the deposit of copper which is formed on the surface of the porous jar. This objection has been entirely removed by the invention of a cell known as the gravity cell, in which the two

FIG. 51.—THE GRAVITY CELL.

liquids are separated from each other by means of their difference of density.

The gravity cell is shown in Fig. 51. The copper is made the lower plate, and is kept covered by a saturated solution of copper sulphate, to insure which a large excess of undissolved crystals of copper sulphate are left in the bottom of the jar covering the copper plate. The zinc plate is suspended above the copper plate in the form of an open wheel, by the means shown. When in action a sharp line of demarkation appears between the denser blue solution of copper sulphate and the less dense clear solution of dilute sulphuric acid, or hydrogen sulphate.

When zinc sulphate is employed instead of dilute sulphuric acid, the electromotive force is somewhat lower than that of the ordinary Daniell cell, but the constancy of the cell is greater.

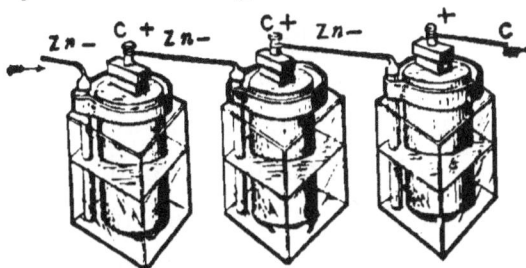

FIG. 52.—THE LECLANCHE CELL.

A form of double-fluid cell of equal importance to the Daniell cell, called the Leclanché cell, is shown in Fig. 52, where three such cells are connected together to form a series battery. The voltaic couple is formed of zinc and carbon. The zinc is immersed in an electrolyte consisting of a dilute solution of sal-ammoniac, while the carbon is surrounded by black oxide of manganese in a finely divided state.

The carbon element consists of a plate or rod of carbon placed inside a porous cell, which contains a mixture of broken gas retort carbon and finely divided black oxide of manganese. The black oxide of manganese here takes the place of the other liquid in the double-fluid cell, since it acts as the oxidizing substance which covers the negative plate, and prevents it from becoming coated with hydrogen by oxidizing or removing such hydrogen. The Leclanché cell gives an electromotive force of about 1.47 volts. It readily polarizes, and is, therefore, capable of furnishing a constant current for but a comparatively short time. It possesses, however, the power of depolarizing if left on open circuit, and, for this reason, is generally called an open-circuited battery.

Of all the voltaic cells that have been devised, two only, namely, the gravity and the Leclanché, have survived in the struggle for existence, and come extensively into general use. The gravity cell is used on what are called closed-circuited lines, and the Leclanché cell on what are called open-circuited lines.

The gravity cell is suitable for all purposes that require a constant current for an indefinite duration of time, such, for example, as in most of the systems of telegraphy operated in the United States, while the Leclanché cell is suitable for such purposes as require a momentary current for ringing bells, the operation of annunciators or for other similar work, and are left on open circuit most of the time.

, A form of voltaic cell convenient for some purposes is found in what is called the dry cell. Such a cell is shown in Fig. 53. The term dry cell is a misnomer, since all such cells are moistened with

FIG. 53.—DRY CELL.

solutions of electrolytes, and are, therefore, far from dry. The moistening is obtained by the use of some hydroscopic substance that absorbs moisture from the air.

The mistake is very commonly made of calling a voltaic cell a voltaic battery. Strictly speaking, a voltaic battery, like any other form of battery, con-

sists of such a combination of a number of separate
cells as will permit them to act as a single cell.

A convenient form of voltaic battery is shown in
Fig. 54. When it is desired to obtain a current, the

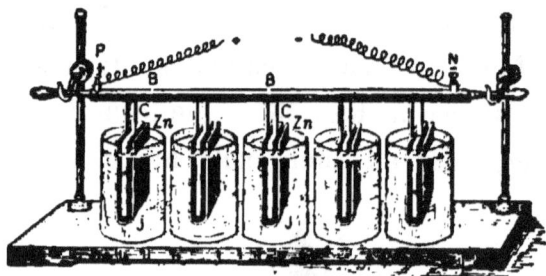

FIG. 54.—PLUNGE BATTERY.

battery plates are lowered into the acid solution in
the cups; when the battery is no longer required for
use, the plates are raised from these liquids.

EXTRACTS FROM STANDARD WORKS.

Larden, in his "Electricity for Public Schools and Colleges,"* page 170, gives the following statement as to the views held by the advocates of the contact and of the chemical theory of the origin of the difference of potential in the voltaic cell :

Volta fixed his attention mainly on the $\triangle V$(difference of potential) that, as it seemed, accompanied the contact of the dissimilar metals *zinc* and *copper*. His followers exaggerated a certain one-sidedness that existed in his views ; and the *Contact school* as they were called, considered that the chemical solution of the zinc played a subordinate part in the action of the cell, serving mainly to keep the surfaces clean and so to keep the same series of bodies in contact. In fact the word *contact* was the keynote to their theory of the voltaic cell. They considered the $\triangle V$ between the terminals of the "open" cell (that is, of a cell in which the terminals were insulated) as the algebraic sum of the different $\triangle V$'s due to the different contacts; of which, in the ordinary Volta's cell, the only one of importance was that where the copper wire was soldered to the zinc.

The "Chemical school" of physicists considered the cell when the circuit was closed and a current was running.

---

* "Electricity for Public Schools and Colleges," by W. Larden, M.A. London : Longmans, Green & Co. 1887. 476 pages, 219 illustrations. Price, $1.75.

They pointed out how the strength of the current that flowed was proportional to the vigour with which the chemical action proceeded ; and how the power of the cell depended on having one plate as much acted upon, and the other plates as little acted upon, as possible.

Faraday was the great exponent of this view.   In modern phraseology, the " Chemical school " insisted on the *chemical a*[*]*tion* as the *source of the energy* of the cell.

They, in their turn, were for the most part too one-sided ; and many denied that dissimilar metals in contact did exhibit a difference of potential at all without chemical action.

In the next section we shall attempt to show the position of modern theory and of modern knowledge in this matter ; and shall conclude by giving a view of the Volta's cell, taken as a whole, which can hardly involve any serious error.

But we should add that the whole question is still to a considerable extent unsettled.

V.—THERMO-ELECTRIC CELLS AND OTHER ELECTRIC SOURCES.

In 1821 Seebeck, of Berlin, discovered another source of electricity in the unequal heating of dis similar metals. He found, when two dissimilar metals are formed into a circuit by soldering their junctions together, that when òne of the junctions is heated above the temperature of the other junction a current of electricity is produced, which flows through the circuit in a certain direction, and that, when this junction is cooled below the temperature of the other junction, the current so produced flows in the opposite direction.

He called such currents thermo-electric currents and the electricity produced by them thermo-electricity.

The two metals or other substances forming a thermo-electric combination are called a thermo-electric couple, and each of the substances forming such a couple, the thermo-electric element.

Thermo-electric phenomena also occur at the junctions of two dissimilar liquids, or at the junction of a liquid and a metal, when such junctions are unequally heated.

One of the simplest ways in which the production of thermo-electric currents can be shown is by soldering two different metals together at one end and connecting their other ends to the terminals of a galvanometer. When the junction is either heated or cooled a current of electricity is produced which deflects the needle of the galvanometer. This deflection occurs in one direction when the junction is heated, and in the opposite direction when it is cooled.

In the following table the different metals are arranged in such an order that each metal acquires positive electricity when combined with any metal following it, and negative electricity when combined with any metal preceding it. Or, in other words, when the junction of two such metals is heated the current passes across the junction through the solder from the +, or positive metal, to the —, or negative metal. Such a series of metals is called a thermo-electric series. A thermo-electric series is given in the following table :

Bismuth	+25	Gas coke	— 0.1
Cobalt	+ 9	Zinc	— 0.2
Potassium	+ 5.5	Cadmium	— 0.3
Nickel	+ 5	Strontium	— 2.0
Sodium	+ 3	Arsenic	— 3.8
Lead	+ 1.03	Iron	— 5.2
Tin	+ 1	Red phosphorus	— 9.6
Copper	} + 1	Antimony	— 9.8
Silver	} + 1	Tellurium	—179 9
Platinum	+ 0.7	Selenium	—290.0

—Ganot.

For example, in the bismuth-antimony couple, when the junction is heated, the current passes from the bismuth, the positive metal, across the junction to the antimony, the negative metal.

The meaning of the numbers is as follows : calling the electromotive force of a copper-silver couple unity, the electromotive force of any other pair where the signs are the same is equal to the difference of the numbers, but where the signs are different is equal to their sum. For example, the electromotive force of the bismuth-nickel couple is 25 — 5=20 times that of the silver-copper couple ; that of the bismuth-antimony couple, $25 + 9.8 = 34.8$.

For small differences of temperature the electromotive forces produced are proportional to the temperature. As the temperature increases the electromotive force decreases until, at a certain temperature of the hot junction, called the neutral temperature, no difference of potential is produced.

A thermo-electric couple joined in a circuit so as to produce electricity is called a thermo-electric cell. A number of thermo- electric cells, so arranged as to act as a single source or cell, is called a thermo-electric battery.

The difference of potential produced by any thermo-electric couple is approximately proportional to

the difference of temperature of its junctions, provided such difference of temperature is not too
great. The actual difference of potential produced
by the best thermo-electric couples is quite low. In
the case of one of the most powerful of such
couples, namely, that of bismuth-antimony, the
electromotive force or difference of potential for one
degree centigrade difference of temperature is only
117 micro-volts. In order, therefore, to increase

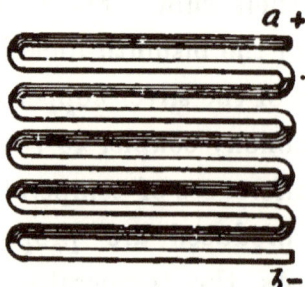

FIG. 55—SERIES-CONNECTED THERMO-ELECTRIC COUPLES.

the difference of potential, a number of separate
thermo-electric cells are joined together so as to
form a thermo-electric battery.

The manner in which the separate thermo-electric
cells are connected in series to form a thermo-electric battery is shown in Fig. 55, which represents
a thermo-electric battery known as Nobilli's thermo-
electric pile, or battery, after the name of its inventor.

Here a number of bismuth-antimony couples are carefully insulated from one another in all parts of their circuits except at their junctions, which are soldered together and are then piled so as to form the cubical pile, shown in Fig. 56.

The different couples in this pile are connected throughout in series, so that the difference of potential produced increases in the direct proportion of the number of separate cells connected together.

FIG. 56.—NOBILLI'S THERMO ELECTRIC PILE.

The battery, or thermo-electric pile, is placed in a metallic box, and the free terminals of the first and last couples are connected to the binding posts to form the terminals of the pile. These binding posts are seen in the figure at the top of the pile.

If the separate junctions of the thermo-electric couples be numbered successively from the first to the last it will be seen that when they are arranged as shown in Fig. 56, all the even junctions are situated at one face of the pile and all the odd junctions at its opposite face. This is necessary in order that the separate thermo-electric differences of potential generated by the separate couples shall be added together ; for, if a thermo-electric circuit be formed as shown in Fig. 57 and all its junctions are equally heated, no thermo-electric currents will be produced, since the differences of potential thereby generated neutralize one another. It is not actually necessary to employ different metals to form

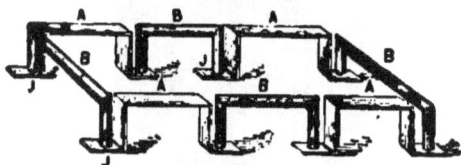

FIG. 57.—THERMO-ELECTRIC CIRCUIT.

thermo-electric couples, since couples formed of even the same metals, under different physical conditions, will produce thermo-electric currents when unequally heated. Such couples, however, are generally very weak.

For example, if a conductor such as a wire, a part of whose length is bent on itself, and the remainder of which is straight, as shown in Fig. 58, be heated at the straight part by the flame *F*, of a lamp, a difference of potential will be produced, as can be shown by connecting the ends of the wire with a galvanometer.

Thermo-piles have been constructed by Clamond of couples of iron and an alloy of zinc and antimony of sufficient power to sustain a voltaic arc producing a light equal to 40 carcel burners. Many practical difficulties exist, however, which will have to be overcome before thermo-piles can be commercially employed as electric sources.

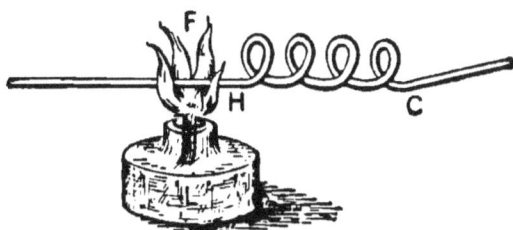

FIG. 58.—THERMO-ELECTRICITY.

It is, however, in the production of electricity directly from heat produced by burning coal that the greatest advance in the science of electricity is to be expected in the near future, and it is possibly in the direction of thermo electric piles that such advance is to be effected.

In thermo-electric piles, considered as sources of electricity, although the contact of dissimilar metals is necessary for such production, yet the source of the energy which must be expended to maintain such current is undoubtedly to be found in the heat energy, for the heat applied at the heated junctions disappears more rapidly when the circuit is closed than when it is opened.

A variety of electric cell, which depends for its source of energy on light, or luminous radiant energy, rather than on heat, or non-luminous radiant energy, is to be found in the photo-electric cell.

Photo-electric cells are made in a variety of forms and of a number of different materials ; selenium, however, a comparatively rare substance, generally found associated with sulphur, is most frequently employed for this purpose.

One of the simplest forms of selenium cell consists of a mass of selenium that has been fused between two conducting wires of platinized silver or other suitable material. The platinized silver wire is wound in two separate spirals around a cylinder of hard wood, care being taken to keep the two wires a constant distance apart so as to avoid any direct contact between them.

The space between these two parallel wires is then filled with fused selenium which is allowed to cool gradually.

This construction, which permits the wires to act as the terminals of an extended plate of selenium, is necessary in order to decrease the resistance of the cell so formed, the electrical conductivity of selenium being very poor.

Such a cell forms what is sometimes called a selenium resistance, and, when its opposite faces are unequally exposed to sunlight, so that one is illumined and the other is kept dark, a difference of potential is thereby produced which will result in an electric current, if the terminals of the cell are connected by means of a conductor.

As in the case of the thermo-electric cell the current flows in one direction if one face is illumined more than the other face, and in the opposite direction if this face be illumined less than the other.

Exposure to sunlight reduces the resistance of a selenium cell to about one-half its resistance in the dark, but such change of resistance does not remain constant for a long time.

A number of curious applications have been made of the currents of electricity produced by selenium cells. The following are examples :

(1.) A selenium cell is so placed in a circuit containing an electro-magnet and switch that, on one of its electrodes being exposed to the decreased illumination of coming night, the current produced automatically turns on an electric lamp, and, conversely, on the approach of daylight, and the consequent illumination of the electrode. turns it off.

(2.) A device has been proposed whereby the presence of a light—as, for example, that carried by a burglar—automatically rings an alarm, and thus calls the attention of a watchman in the building.

A selenium cell is employed in connection with a variety of apparatus as a resistance that is automatically variable on exposure to light. When inserted along with suitable electro-receptive apparatus in the circuit of an electric source—such, for example, as a voltaic battery—on the exposure of one of its faces to the light its resistance decreases and thus permits the passage of a stronger current through the circuit in which it is placed and the consequent energizing of the electro receptive apparatus. The photophone, a variety of telephone, is constructed on this principle.

A device called the selenium eye is also constructed on the same principle. In this apparatus a diaphragm, the aperture of which represents the pupil of the eye, is automatically dilated and contracted by means of light, which falls on a selenium resistance.

Such an apparatus is shown in Fig. 59. A selenium cell *S,* placed as shown, is provided with two slides or lids *L, L.* When these are moved toward each other the amount of light falling on the selenium resistance is decreased ; when they are moved in the opposite direction such amount is increased.

This motion may be obtained automatically by inserting in the circuit of a voltaic battery an electro-magnet, the movements of whose armature draw the slots together, and the movement of a spring

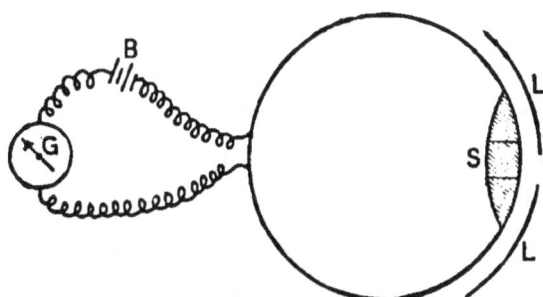

FIG. 59.—SELENIUM EYE.

moves them in opposite directions on the weakening or cessation of the current. When, now, light falls on the selenium resistance *S,* the resistance of the circuit is decreased, and the movement of the arma- ture of the electro-magnet thereupon draws the slots together, thus decreasing the amount of light.

Van Uljanin constructed a selenium cell as follows : Selenium was melted in between two parallel platinized silver wires. On being gradually cooled under pressure, after heating up to 195° C. in a paraffine bath, and this being repeated several times, the selenium was changed from the amorphous to the sensitive, crystalline variety.

By the employment of such a cell Van Uljanin established the following general principles :

(1.) On exposure to light an electromotive force is developed which produces a current flowing from the dark or non-illumined electrode to the illumined electrode.

(2.) The greatest electromotive force, which is 0.12 volt, disappears instantaneously and completely on the removal of the light ; or, in other words, the action of the light is practically instantaneous.

(3.) The resistance and sensitiveness to light, as well as the production of the electromotive force, decreases with the age of the cell. This is probably due to a gradual change in the allotropic state in the selenium.

(4.) The electromotive force is proportional to the intensity of the illumination only when the obscure rays or heat rays are absent.

Besides the photo-electric cell already referred to, differences of temperature in certain crystals are also able to produce electricity. When a crystal of tourmaline, or other pyro-electric substance, is heated or cooled, it acquires electrification at points called its poles.

In the crystal of tourmaline shown in Fig. 60 the end *A*, called the analogous pole, acquires a positive electrification, and the end *B*, called the antilogous pole, negative electrification, while the temperature of the crystal is rising. While cooling the opposite electrifications are produced.

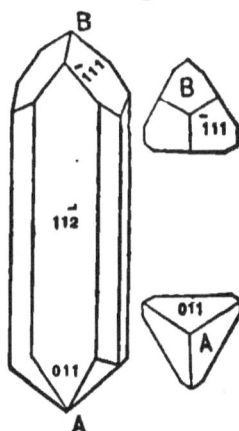

FIG. 60.—PYRO-ELECTRIC CRYSTAL.

A heated crystal of tourmaline, suspended by a fibre, is attracted or repelled by an electrified body, or by a second heated tourmaline, in the same manner as an electrified body.

Many crystalline bodies possess similar properties. Among these are boracite, quartz, tartrate of potash, sulphate of quinine and an ore of zinc known for this reason as electric calamine.

Another curious source of electricity is to be found in the currents produced when liquids are forced to pass through the capillary spaces in thin walls.

In the form of capillary electrometer shown in Fig. 61 the horizontal glass tube *B*, filled with mercury, has a drop of sulphuric acid at *B*. Its open ends are connected with two vessels *M* and *N*, also filled with mercury. When an electric current is passed through the tube a movement will be observed in the drop of sulphuric acid in a direction

FIG. 61.—CAPILLARY ELECTROMETER.

which changes with the change in the direction of flow of current through the tube. Now Quinque has shown that, if such a drop of liquid be moved by mechanical force, an electric current is produced the electromotive force of which will depend :

(1.) On the material of the diaphragm.

(2.) On the nature of the liquid.

(3.) On the pressure required to force the liquid through the diaphragm.

The bodies of plants and animals are active sources of electricity. The exact causes which produce this electricity are as yet unknown, but it is certain that, in one way or another, these electric currents are necessary for the carrying out of the vital processes of the animal or plant.

Electricity, passed through the bodies of either animals or plants, produces convulsive movements therein.

Considerable advantage has been made of such electrical currents for the treatment of diseased conditions of the body. When properly employed, such treatment is of undoubted value in the curing of certain diseases, but the use of so powerful an agency as electricity should not be attempted except by a skilled physician.

Some animals appear to possess certain organs especially designed to produce electric discharges sufficiently powerful to serve as a protection against their enemies. Such is the case with the electric eel, a drawing of which is shown in Fig. 62.

According to Faraday a shock given by a specimen of electric eel which he examined was equal to that produced by 15 Leyden jars having a total surface of metallic coating of 25 square feet.

Du Bois Reymond has shown that during vigorous

growth plants are active sources of electricity. If one of the terminals of a galvanometer be inserted in a fruit near its stem end, and the other terminal in the opposite end of the fruit, the needle of the galvanometer at once shows the pressure of an electric current.

Buff has shown that the roots and the interior

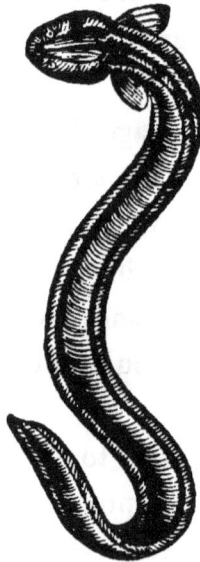

FIG. 62.—ELECTRIC EEL.

portions of plants are always negatively charged, while the flowers, fruits and green twigs are positively charged.

Plant-tissue, or fibre, like the muscular fibre of animals, exhibits in many cases a true contraction on the passage through it of an electric current.

EXTRACTS FROM STANDARD WORKS.

In Noad's "Student's Text-Book of Electricity,"* revised by Preece, the following facts are given concerning some forms of thermo-electric piles, p. 396:

Prof. Dove employed iron and platinum soldered together in alternate lengths, and the whole wound on a cylinder of such diameter as to bring all the iron-platinum junctions on one side of the cylinder, and all the platinum-iron junctions on the other.

Farmer, in America, used Marcus metal and German silver, but failed to secure a good permanent connection.

Instead of bismuth and antimony, Bunsen used as the elements of his thermo-battery copper pyrites combined with copper, or *pyrolusite* combined with copper or platinum. Ten of such combinations give all the actions of a Daniell battery, having an effective copper surface of 14 square centimetres (2.17 square inches) in area. Stefan employed granulated sulphide of lead for the positive, and copper pyrites for the negative element—the power of a single pair, as compared with a Daniell's cell, is stated to be as 1 to 5.5. Marcus, in Germany, used for the positive metal an alloy composed of copper, 10 parts ; zinc, 6 parts ; nickel, 6 parts ; and for the negative an alloy composed of antimony, 1 parts ; zinc, 5 parts ; bismuth, 1 part.

---

* "The Student's Text Book of Electricity," by Henry M. Noad, Ph. D., F. R. S. Revised by W. H. Preece, M. I. C. E. London: Crosby Lockwood & Co. 1879. 615 pages, 471 illustrations. Price $4.

The elements of his battery are about 7 inches long, 7 lines broad, and half a line thick; they are screwed together, and so arranged that their lower junctions can be heated by a row of gas jets, and the upper cooled by a current of water. The electromotive force of one element is equal to one-twenty-fifth of a Bunsen's cell. *Six* pairs decompose water; *thirty* pairs cause an electromagnet to lift 150 pounds; and *one hundred and twenty-six* pairs decompose water at the rate of 25 cubic inches of mixed gases per minute, and melt a platinum wire half a millimetre in thickness.

The conversion of heat into electricity by this battery, is strikingly shown by the fact that the water used for cooling the upper junctions, is more rapidly warmed when the current is broken, than when it is closed.

It was stated by Marcus, in a communication to the Austrian Academy of Sciences, that he had constructed a furnace consuming 240 pounds of coal per day, intended to heat 768 elements of his thermo-battery, the electromotive power of which would be equivalent to 30 cells of Bunsen's nitric acid arrangement.

Wheatstone constructed a thermo-pile on Marcus's principle, the electromotive force of which was equal to *two* Daniell's cells; he found that its power was greatly increased by repeatedly melting the alloy composing the bars, probably in consequence of their crystalline structure being thereby broken down.

VI.—DISTRIBUTION OF ELECTRICITY BY DIRECT OR CONTINUOUS CURRENTS.

The electricity produced by any source or battery of sources is, by means of variously arranged circuits, distributed to electro-receptive devices placed in connection therewith.

Any system for the distribution of electricity consists, therefore, of essentially the same combinations of parts as are found in any circuit ; namely,

(1.) Of various electric sources or batteries of electric sources.

(2.) Of various electro-receptive devices.

(3.) Of conductors or leads connecting the electric sources, or battery of electric sources, with the electro-receptive devices.

The term direct current, as generally employed, signifies a continuous current, or one that flows continuously in the same direction, as distinguished from an alternating current, or one that flows alternately in opposite directions. The term constant current is sometimes applied to the case of a current that is not only continuous in the sense of always flowing in the same direction, but is also such a current as maintains approximately a constant current strength.

A number of different systems have been devised for the distribution of electricity to the electro-receptive devices. Among the most important of these are the following:

(1.) Distribution by means of direct or continuous currents.

(2.) Distribution by means of alternating currents.

(3.) Distribution by means of storage batteries, or secondary generators.

(4.) Distribution by means of condensers.

(5.) Distribution by means of motor-generators.

The most important purposes for which electricity is distributed are for the production of light, power, heat, and for telegraphic or telephonic communication.

Distribution by means of direct or continuous currents may be effected in a variety of ways; they can, however, be arranged under two general classes:

(1.) Means whereby the current is so distributed over the line wire or conductor that its strength shall be maintained approximately constant notwithstanding changes in the number of electro-receptive devices that have been introduced into or removed from the circuit. Such a distribution necessitates the connection of the electro-receptive devices to the circuit in some form of series circuit, and is, therefore, generally called a system of series distribution. It is also sometimes called a system of constant current distribution.

(2.) Means whereby electricity is so distributed over line wires that the potential difference, or electromotive force, will be maintained approximately constant, notwithstanding changes in the number of electro-receptive devices that have been introduced into or removed from the circuit.

Such a distribution necessitates the connection of the electro-receptive devices to the line in some form of multiple connection, and is, therefore, generally called a system of multiple-arc or parallel distribution. It is also sometimes called a system of constant potential distribution.

In a system of series distribution, the electro-receptive devices are connected to the main line in series in the manner shown in Fig. 63, so that the current passes successively through each of the electro-receptive devices.

In a series system of distribution, each electro-receptive device added increases the resistance of the circuit, the total resistance of the circuit being equal to the sum of the separate resistances placed therein. In order, therefore, to maintain the current strength constant, the electromotive force of the source must be increased for each electro-receptive device added, and must decrease for each electro-receptive device removed.

The number of electro-receptive devices which are placed in series in a series distribution circuit is sometimes so great that the electromotive force of the source must be very high. The high poten-

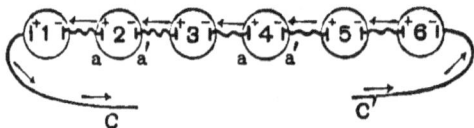

FIG. 63.—SERIES-CONNECTED ELECTRO-RECEPTIVE DEVICES.

tial required is either obtained from a properly proportioned single source or from a suitably connected battery of separate sources.

The commercial requirements of series distribution necessitate the frequent introduction and removal of the electro-receptive devices from the circuit. Means must, therefore, be devised whereby the electromotive force can be automatically varied to meet the requirements of the circuit at any time.

The distribution of arc lamps in series forms one

of the principal purposes for which series distribution is employed.

In systems of arc light distribution the electric
source is almost invariably a dynamo-electric machine, or battery of dynamo-electric machines. The
variations in the electromotive force required to
maintain a constant current in the circuit, despite
changes in the load or the number of electro-
receptive devices in such circuit, are insured by
means of some system of regulation whereby the
electromotive force furnished by the machine can be
altered.

Such regulation may be obtained either automatically or by means of an attendant.

The automatic regulation for a constant current
may be obtained either by shifting the position of
the collecting brushes on the commutator cylinder,
as in the Thomson-Houston system, or by the use of
a variable resistance placed as a shunt to the circuit
of the field magnets of the dynamo, as in the Brush
system.

The means employed in the Thomson-Houston
system of automatic current regulation are shown in
connection with Fig. 64. The collecting brushes
are fixed to levers that are moved by the regulator
magnet R, the armature of which is provided with

an opening for the entrance of the paraboloidal pole piece. In order to prevent too sudden movements of this lever a dash-pot is provided.

The adjustment is such that when the current strength is normal the coil of the regulator magnet is short-circuited by contact points at *S, T,* which act as a shunt of low resistance. These contact points are opened or closed by a solenoidal magnet

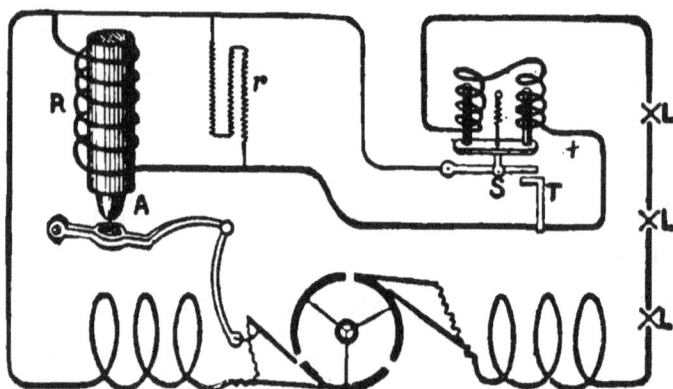

Fɪɢ. 64.—Tʜᴏᴍsᴏɴ-Hᴏᴜsᴛᴏɴ Sʏsᴛᴇᴍ ᴏꜰ Aᴜᴛᴏᴍᴀᴛɪᴄ Rᴇɢᴜʟᴀᴛɪᴏɴ

called the controller, whose coils are placed in the main circuit.

The cores of the solenoidal magnets are suspended by means of a spring. The contact points are opened when the current becomes too strong, and the current traversing the coils of the regulator magnet *A,* attracts its armature, the movement of which shifts

the collecting brushes into a position on the commutator, at which a smaller current is taken off.

In order to decrease the spark that occurs at the
contact points S and T, on the opening of the circuit, a carbon shunt, r, of high resistance is placed
in the circuit in the manner shown.

In actual operation the contact points are continually opening and closing, thus maintaining a practically constant current in the series circuit.

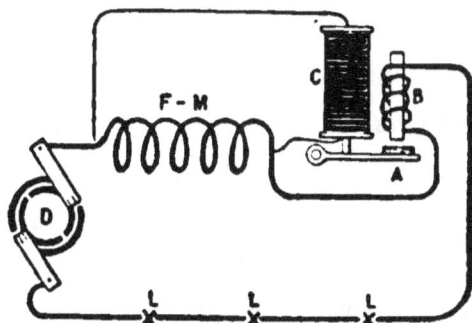

FIG. 65.—BRUSH SYSTEM OF AUTOMATIC REGULATION.

The system of automatic regulation employed by
Brush is shown in Fig. 65. In this system of regulation a resistance C, placed so as to form part of a
shunt circuit to the field magnets of the machine
$F M$, has its value automatically varied. This resistance is formed of a pile of carbon plates, packed,
as shown, in a cylindrical vessel.

On an increase of the current strength—such, for
example, as would result from the extinguishing of

some of the lamps in the circuit—the electro-magnet
B, placed in the main circuit, attracts its armature
A, and thus compressing the carbon plates in *C*,
lowers their resistance. This lowering of the resistance diverts a larger proportion of the current from
the field-magnet   coils *F M*, and maintains the
current strength practically constant.

In some forms of dynamo-electric machines the
regulation is effected by an assistant. This is ob-
jectionable.

In the series distribution of electricity the current
is passed successively through the electro-receptive
devices. In order to prevent the failure of any
single device from opening or breaking the entire
circuit some form of safety device must be employed.
Such devices generally consist of an arrangement by
means of which a short circuit of comparatively low
resistance is automatically established past the faulty
device.

In arc light circuits these safety devices generally
consist of a circuit of low resistance formed of heavy
contact points that are closed by means of an electro-magnet placed in a shunt circuit of high resistance around the electrodes of the lamp. Such a device practically forms an automatic switch. In
such cases the lamp is not actually cut out or re-

moved from the circuit, but only practically so cut
out, since most of the current passes through the
circuit of low resistance.

In order to readily cut out or remove a lamp from
the line by hand, so, for example, as to permit it
to be safely recarboned, a hand switch is provided
by means of which a by-path of low resistance is
closed around the lamp, thus practically cutting it
from the circuit. ˇ

In a system of multiple-arc, or parallel distribution, the electro-receptive devices are connected to

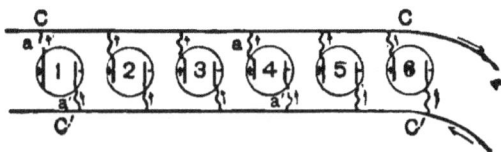

FIG. 66.—MULTIPLE-CONNECTED ELECTRO-RECEPTIVE DEVICES.

the leads or conductors, which constitute the main
line, in some of the varieties of multiple or parallel
connections. Each of the devices added decreases
the resistance of the circuit, since it increases the
area of cross section of the conductors that connect
the opposite leads.

The multiple-arc connection of six electro-receptive devices, 1, 2, 3, 4, 5 and 6, to the leads C, C,
and C', C', is shown in Fig. 66.

In order to maintain constant the strength of the current which passes through each device, notwithstanding a change in the number of such devices, the difference of potential at the terminals of each of the devices added, which are here supposed to be of the same resistance, must likewise be maintained constant.

It will be remembered that in the series distribution circuit the current strength must be maintained constant in order to insure the passage of the same current through each device added to the circuit, no matter how many might be placed therein at any one time.

In the multiple connection, however, it is the potential or the electromotive force of the leads to which the circuit is connected that must be maintained constant. The series-connected circuit, as already mentioned, therefore, is sometimes called a constant-current circuit, and the multiple-connected circuit is sometimes called the constant-potential circuit.

In the constant-current circuit each device added to the line in series increases its resistance. In order to avoid too great a resistance in such a circuit the resistance of each device added is generally comparatively small.

In the constant-potential circuit, on the contrary, each device placed in multiple between the leads decreases the resistance of the circuit. In order to avoid too great a decrease in the resistance of such a circuit, where the number of receptive devices is great, the resistance of each separate device is generally high.

The constant difference of potential required on the leads of a constant-potential circuit where dynamos are employed as the electric source is generally obtained either by some form of compound-winding or by means of hand regulation, or by both.

An automatic regulation by means of compound-winding of dynamos is particularly applicable to constant-potential machines. By compound-winding, the magnetizing effects of the shunt coils is maintained approximately constant, while that of the series coils varies in proportion to the load that is on the machine.

In compound-wound machines the series coils are sometimes wound close to the poles of the machine and the shunt coils nearer to the yoke of the magnets, though custom varies somewhat in this respect.

The object of compound-winding is to render the dynamo self-regulating under changes in its work-

ing load. In the compound-wound dynamo, the
shunt coils are often, for convenience, superposed
on the series coils and consist of a much greater
number of convolutions of fine wire than is placed
in the series coils, which are of coarse wire.

Suppose, for example, the terminals of the shunt
coils of a compound-wound dynamo are connected
to the binding posts of the machine. When the
current leaves the armature it has two paths, one

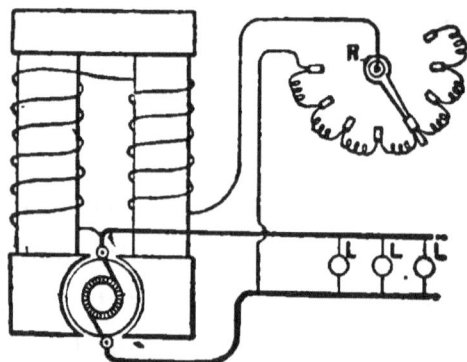

FIG. 67.—HAND REGULATION.

through the thick series coils to the external circuit,
and the other through the finer and longer shunt
coils. The resistance of the shunt coils is so much
greater than that of the armature that the current
variations in the armature will produce no appreciable effect on the magnetizing power of the shunt,
which, therefore, acts as a nearly uniform exciter of
the field.

The hand regulator shown in **Fig. 67** is a device employed on some of the Edison dynamo-electric machines. The variable resistance at *R*, connected to the machine and one of the leads as shown, is provided with a lever-switch, which is operated by hand whenever the potential rises above or falls below its proper value. By these means an approximately constant potential is maintained on the leads to which the lamps *L*, *L*, *L*, are connected in multiple.

In systems of multiple connection, since the electro-receptive devices are connected to the leads in multiple, the opening of the circuit of any single device does not interfere with the operation of the remainder of the devices placed therein. It is necessary, however, in practice, for the purpose of preventing abnormally great currents from passing through the circuit of any single device, and thereby destroying the balance of the rest of the circuit, or raising the temperature of the circuit dangerously high, to employ devices called **safety-fuses**, strips or plugs. These consist essentially of strips, plates or bars of lead, or some other readily fusible alloy, which are placed directly in the circuit, and which fuse and automatically break the circuit on the passage of any current that would injure the safety of other parts of the circuit.

These safety-fuses are placed both in the branch circuits and in the main line circuits. Fig. 68 illustrates such a safety-fuse, as arranged for a cutout.

FIG. 68.—CUT-OUT.

Instead of the distribution of incandescent lamps by multiple connection a multiple-series connection is often employed· because it permits of a higher difference of potential being maintained on the leads.

The system of distribution known as the three-

wire system is in reality a modification of the multiple-series distribution.

The lamps or other electro-receptive devices are
placed in multiple-arc between either branch, and
are distributed so that the current in each branch
is approximately the same.

No current passes through the central conductor
when the balance is established, but when the

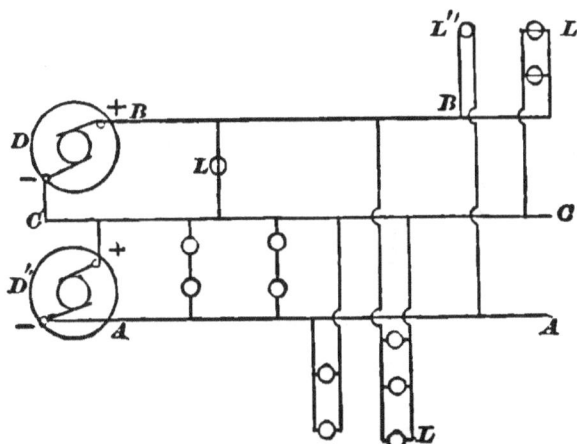

FIG. 69.—THREE-WIRE SYSTEM.

balance is destroyed this central conductor takes up
such surplus current.

In the three-wire system double the usual difference of potential is used that is required for a single
lamp, and a considerable saving is thereby effected
in the cost of the leads or mains used therewith.

The arrangement of the parts in a three-wire

system of distribution are shown in Fig. 69. The
dynamo *D*, has its negative and *D'*, its positive pole
connected to the conductor *C' C*, called the neutral
conductor. The dynamo *D'*, has its free negative
terminal connected to the negative lead *A A*, and
the dynamo *D*, has its free terminal, the positive terminal, connected to the positive lead *B B*.
The lamps are connected as shown at *L, L, L,
L"*.

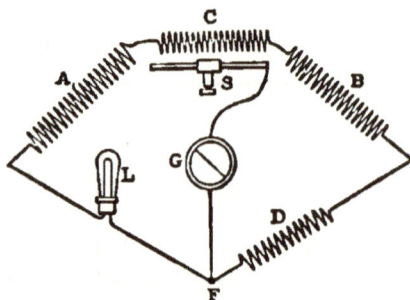

FIG. 70.—EDISON-HOWELL LAMP INDICATOR.

An examination of the drawing will show that
if desired the entire difference of potential generated by the two dynamos can be fed to a single
electro-receptive device as shown at *L"*, or only
the difference of potential between the neutral wire
and the other leads can be utilized, as at *L, L*.

In any system of incandescent lamp distribution
in multiple it is necessary to know at the central
station whether or not the proper voltage or differ-

ence of potential exists on the mains. An apparatus for this purpose, shown in Fig. 70, is known as
the Edison-Howell lamp indicator.

The apparatus depends for its operation on the
variations which are produced in a carbon resistance
consequent on changes in its temperature. The resistance employed for the purpose is a carbon incandescent lamp, such as is employed in the ordinary
commercial circuits. The apparatus consists essentially of a Wheatstone bridge with resistances arranged as shown. A galvanometer at G, serves by
the movements of its magnetic needle as an indicator. Its needle remains at zero as long as the potential difference has the exact voltage required on the
circuit with which the indicator is connected, but
moves to one side or the other whenever an increase
or a decrease occurs in the potential difference.

The incandescent lamp at L, which is one of the
resistances, and is constantly traversed by the current, will have a fixed resistance for the temperature
at which it is designed to run.

The other resistances are so proportioned as to
insure the needle at G, remaining at zero. If,
however, the potential varies, the temperature of the
lamp L, varies, its resistance also varies and, being
carbon, there is a rise of temperature corresponding

to a fall of lamp resistance, which destroys the balance of the bridge and deflects the galvanometer needle. The attendant then regulates the potential to bring the needle back to zero.

The necessity for an efficient lamp indicator will be understood when it is remembered that it is necessary in the commercial use of incandescent lamps to maintain as nearly a constant potential on the mains as possible. A decrease in potential will result in a decrease in the candle power. An increase in potential, although attended by an increase in the candle power, produces a marked decrease in the life of the lamp, such decrease following an increase of but a few per cent. in the difference of potential.

EXTRACTS FROM STANDARD WORKS.

The following statements concerning current distribution are made by Slingo and Brooker on page 579 of their *"Electrical Engineering."**

In systems of distribution of electrical power by means of constant current the question is comparatively simple, as the current employed is not a heavy one, and has the same value at all times and in all parts of the circuit. The chief difficulty likely to arise is in providing for future extensions of the system when the potential difference which can be applied at the ends of the circuit is limited. The more interesting and more difficult problem consists in the supply of currents to lamps, or other apparatus, at a constant potential; for then the main conductors have to carry a very heavy and variable current. The matter becomes more difficult if the lamps are distributed over a wide area, or are situated at a distance from the generating station. As has been pointed out in chapter XIII., the power wasted may in such cases be reduced to a minimum by transmitting it in the form of a small current at high pressure, and reducing the pressure at the required point to a suitable value. But such a system has its disadvantages. Although the cost of the copper is vastly reduced, the high potential difference employed demands very efficient and expensive

* "Electrical Engineering for Electric Light Artisans and Students," by W. Slingo and A. Brooker. London : Longmans, Green & Co. 1890. 631 pages. 307 illustrations. Price $3.50.

insulation, the engines and dynamos must always be kept running, and when very little power is being demanded the efficiency of the transformers and the whole system falls to a low value. For even when the secondary circuit of a parallel transformer is disconnected, some current passes through the primary, and when only one or two lamps are joined up, the power appearing in the secondary may be but a comparatively small fraction of that absorbed by the primary. When the number of transformers is large, the total power wasted becomes considerable during the time when little or no light is required.

In the other method of distributing direct from the dynamo to a number of lamps all joined up in parallel, the chief problems to be faced are the heavy loss occurring in the mains and the difficulty of regulating the supply to each lamp.

VII.—ARC LIGHTING.

A comparatively short time after the invention by Volta of the voltaic pile, Sir Humphry Davy, by means of a powerful voltaic pile of 2,000 couples, showed at the Royal Institution the full splendors of the voltaic arc. Although this was not the first time that light was obtained from the carbon arc, yet it was the first time it was publicly shown on so extended a scale, and the exhibition practically led to electric arc lighting, now so generally employed for the illumination of extended areas.

When the terminals of a sufficiently powerful dynamo or other electric source are connected to two carbon pencils or rods that are first placed in contact and afterward gradually separated, a brilliant arc or bow of light appears between them. This is called the voltaic arc, from Volta, the discoverer of the battery by the use of which it was first obtained. It takes the name arc from its arc or bow shape.

In order to form a voltaic arc, the carbon or other electrodes are first placed together and then gradually separated. The arc which is formed between them consists mainly of volatilized carbon. A part of the current raises the carbon to high incandescence, and, when the carbons are gradually separated, the current flows or passes from one carbon through the mass of glowing vapor to the other carbon.

The volatilization of the carbon forms a tiny depression or crater at the point where the current leaves one carbon and a tiny projection or nipple on the other carbon, formed by the deposition of that portion of the carbon vapor which is not consumed by combustion. This nipple consists of almost pure graphite.

When a voltaic arc is formed between metallic electrodes, a flaming arc is obtained the color of which is characteristic of the burning metal; thus copper forms a brilliant green arc. The metallic arc, as a rule, is much longer and less brilliant than an arc with the same current taken between carbon electrodes.

The light-giving power of a heated body increases very rapidly with its temperature. This increase in light is believed to be as great as the sixth power of the increase in temperature; that is to say, if the temperature of any body is doubled, its power of emitting light will be increased 64 times, or as 2^6. Although the voltaic arc, as well as the carbon electrodes, is intensely heated, yet the greater part of the light emitted comes from the tiny crater in the positive carbon. When, therefore, the voltaic arc is employed as a source of light, as in the arc lamp, and such light is desired to be turned on the space below the lamp, care should be taken, in all forms of lamps where the carbons are supported vertically one above the other, that the upper shall be the positive carbon.

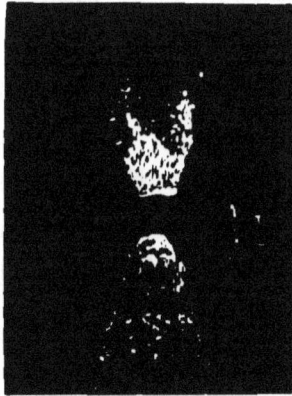

FIG. 71.—VOLTAIC ARC.

During the production of the voltaic arc both positive and negative carbons are consumed by gradual burning; the positive carbon, however, is also consumed by volatilization. The rate of consumption of the positive carbon is, therefore, greater than that of the negative carbon.

The general appearance presented by the carbons
after the voltaic arc has been established between
them for some time is shown in Fig. 71.   Here the
crater in the upper or positive carbon can be seen at
its extreme end, as also the nipple situated at the
opposing end of the negative carbon.   The rounded
globules that are seen on the surface of both carbons
are caused by deposits of molten matters which occur as impurities in the carbons.   In order to employ
the voltaic arc as a source of artificial illumination,
it is necessary to maintain the carbons at a constant
distance apart during their consumption.   This is
accomplished by means of various devices called arc
lamps, which consist essentially of means by which
the carbons are automatically maintained a constant
distance apart during their consumption.

The carbons may be placed in various positions in
arc lights ; namely, either parallel, horizontal, in-
clined or vertically above one another.   The latter
disposition is the one generally adopted.

An arc lamp consists essentially of the following
parts :

(1.) Of various feeding devices for maintaining
the carbons at a constant distance apart during their
consumption.

(2.) Of carbon holders for holding the carbon

pencils, connected with metallic rods called the lamp-rods.

(3.) Of various clutching or clamping devices which grip or hold the lamp rod, and are automatically released by the action of electro-magnets when the length of the arc has exceeded a certain limit.

In most forms of lamps when the lamp is not in operation the carbons are in contact with one another. On the passage of the current, an electro-magnet through whose coils the direct current passes, separates them a short distance from one another by the movement of its armature, and thus establishes an arc between them.

Various forms have been given to the feeding devices of arc lamps. Where the positive carbon is placed vertically above the negative carbon its motion toward it is accomplished by the action of gravity. The lamp-rod being held in a fixed position by the action of a clutch or clamp, the carbon pencil connected with the lamp-rod is held at a certain distance from the fixed negative carbon.

When, now, by consumption, the space or interval between the carbons becomes greater, an electro-magnet, whose coils are placed in a shunt circuit of high resistance around the electrodes, automatically releases the clutch or clamp when a certain distance between the carbons has been reached, and permits the upper carbon to fall toward the lower carbon. In a well constructed lamp, however, the carbons never touch each other, the same electro-magnetic device which releases the clamp automatically clamping it again as soon as the upper carbon by its fall decreases its distance from the lower carbon by the amount desired.

The automatic operation of this shunt magnet will be understood from the following considerations: The resistance of the shunt magnet coils is so much higher than the resistance of the voltaic arc that when the carbons are the proper distance apart too small a current strength flows in this part of the circuit to permit the pull of the armature of the electro-magnet placed in this circuit to release the clutch and permit the fall of the lamp-rod. When, however, by the consumption of the carbons, the distance between them increases, and consequently the resistance of the arc increases, a stronger current flows through the coils of the shunt magnet, and, when such lengthening of the arc has reached a predetermined limit, the increased current, flowing through the coils of the shunt magnet, becomes sufficiently strong to release the clutch or clamp, and thus permit the feeding of the upper carbon. In a well constructed lamp such feeding is almost imperceptible.

Arc lamps are generally placed in series circuits ; that is, in circuits in which the current passes successively through all the lamps in the circuit and returns to the source. In order to avoid the breaking of the entire circuit through the extinguishing of a single arc, an automatic safety device is provided for each lamp, which consists essentially in an electro-magnet so placed in a shunt circuit around the arc that, as the resistance of the arc becomes too great, the increased current, which will then flow through the coils of the magnet, will produce a movement of its armature which closes a short circuit around the lamp, and thus cuts it out of the circuit.

Arc lamps are made in a variety of forms. A well-known form is shown in Fig. 72. The space at the top of the lamp is provided for the movement of the lamp-rod. The feeding mechanism is placed in the cylindrical box near the top of the lamp frame. The open globe is placed around the arc to shield it from the direct action of the wind, as well as for the purpose of scattering or diffusing the light. For this latter purpose the globe is generally ground.

A certain limit of length of the carbon electrodes employed in arc lighting is soon reached in actual practice in all forms of lamps where the electrodes are placed vertically one above the other. Since the lower end of the upper carbon is placed near the upper end of the lower carbon, at the beginning of the consumption the upper carbon and the long

FIG. 72.—ARC LAMP.

lamp-rod connected therewith must necessarily extend considerably above the point where the arc appears. When, therefore, such a lamp is required to be hung near the ceiling of a room, the maximum length which can be given to such a rod is necessarily limited. It has been found in actual practice with

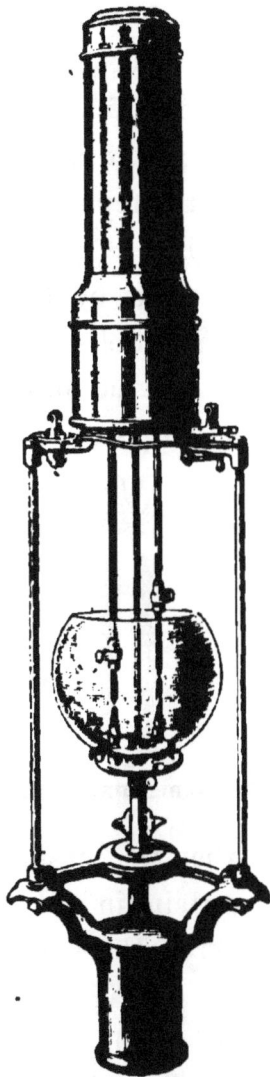

FIG. 73.—ALL-NIGHT ARC LAMP.

the length of carbon rods usually employed, that
when it is desired to maintain the light during the

entire time of darkness, from sunset to sunrise, that one pair of carbons will be consumed and will require to be renewed some time during the run. In the early history of arc lighting it was customary to recarbon the lamps, or to replace the carbons during the middle of the night. A great improvement in this respect has been made by the device called the double-carbon, or all-night electric lamp.

In the all-night electric lamp, two pairs of carbon rods are placed in the lamp, and so connected with the

FIG. 74.—JABLOCHKOFF CANDLE.

lamp circuit that when the consumption of the first pair has reached a certain limit, the current is automatically shifted over to the second pair. Such a form of lamp is shown in Fig. 73. ·

In an early form of arc lamp, called the Jablochkoff candle, two candles placed parallel to each other are maintained at a constant distance apart by some insulating material, such as kaolin, placed between them, as shown in Fig. 74. The current passing into and out of the lamp at one end of the candle forms a voltaic arc at the other end. As the carbons are prevented from moving together by the insulating material placed between them, it is necessary to start the arc by means of a small strip formed of a mixture of some readily ignitable substance, called the igniter, placed between the carbons at the upper end of the candle.

Although the arc starts at the top of the candle when the carbons are of the same length, it can be readily seen that the Jablochkoff candle cannot be used with a direct or continuous current, since, although at the start the carbons are of the same length, yet the more rapid consumption of the positive carbon would soon cause its end to fall so much below the level of the negative carbon as to cause the extinguishment of the light from the too great distance or interval between them. For this reason the Jablochkoff candle is used with alternating currents.

The Jablochkoff candle was at one time extensively employed in arc lighting, but it has been found in practice that the ordinary arc lamp gives more economical results. It is a well-known fact, however, that the Jablochkoff candle gives a much more pleasant and steadier light than most forms of arc lamps. The cause of this, which is somewhat curious, is unquestionably to be found in the fact that in a Jablochkoff candle the light is practically extinguished as many times per second as the alternating current changes its direction. These changes in direction,

FIG. 75.—ARC LAMP HOOD.

however, occur so frequently that before the effect produced on the eye by one change has passed, the next effect follows it, and an average effect is produced which gives the impression of a constant light, the smaller changes of intensity that occur during the existence of any one of these arcs being so much less than that produced by the practically total extinguishment of the arc, that the eye fails to appreciate them. It will be understood, of course, that, in order to prevent the successive extinguishments of the arc from producing perceptible variations in the intensity of the light, they must follow one another with great rapidity.

For the double purpose of protecting the body of the lamp from the rain and sun, and for throwing

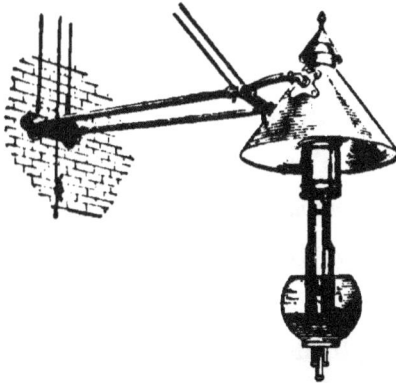

FIG. 76.—OUTRIGGER AND HOOD.

the light downward, a conical shaped hood is provided for all lamps exposed to the weather. Such a hood is shown in Fig. 75.

These hoods are placed either directly on the top of the poles that hold the lamp and circuit wires, or are suspended to the same by suitable supports. When it is desired to suspend such a hood from the side of a building, a special form of support called an outrigger is provided. Such a form is shown in Fig. 76.

The carbon electrodes employed in the early history of arc lighting were made directly from the deposits of carbon that were left in the interior of the gas retorts employed for the production of illuminating gas by the destructive distillation of coal. They are now made by the following process :

Powdered coke, or gas retort carbon, sometimes mixed with lamp-black or charcoal, is made into a stiff dough with molasses, tar, or some other hydrocarbon liquid. The mixture is molded into rods, pencils, plates, bars or other desired shapes by the pressure of a powerful hydraulic press. After drying, the carbons are placed in crucibles and covered with lamp-black or powdered plumbago, and raised to an intense heat, at which they are maintained for several hours. By the carbonization of the hydro-carbon liquid the carbon paste becomes strongly coherent, and, at the same time, by the action of heat, the conducting power of the carbon increases.

To give increased density after baking, and so prolong the life of the pencils, they are sometimes soaked in a hydro-carbon liquid, and subjected to rebaking. This may be repeated a number of times.

Carbons for arc lights are generally copper coated, so as to insure a smaller resistance, a more uniform consumption, and a better contact at the holders.

The unsteadiness sometimes noticed in arc lights is due to a variety of causes, the principal of which are the following :

(1.) Unsteadiness in the driving power, either arising from variations in the amount of power, or from the slipping of the driving belt.

(2.) Imperfections in the working of the feeding mechanism of the lamp.

(3.) Impurities in the carbon.

Unless the carbon electrodes are carefully manufactured they will contain minute portions of materials that are more readily volatilized than those forming the main body of the carbon. When such softer portions are reached, their sudden volatilization a marked variation in the intensity of the light.

Much of the unsteadiness of the arc light is due to the traveling of the arc from one side of the carbon pencil to the other, so that an observer on one side of the lamp will at one moment see the intense light caused by the arc appearing on the side nearest to him, and at the next moment will be exposed to a much less brilliant light by the arc moving to the side furthest from him. This is especially the case

when the carbon pencils are made too thick. It has
been avoided to a certain extent in some cases by
employing carbon electrodes provided with a central
core of charcoal or other softer carbon, which maintains the arc centrally between the electrodes.

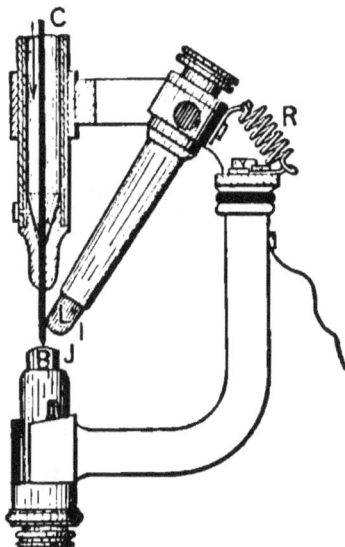

FIG. 77.—SEMI-INCANDESCENT LAMP.

In the form of lamp shown in Fig. 77 the source
of light is due both to the arc established at J, between the lower end of the carbon and the contact
block B, as well as to the incandescence of the pencil or rod of carbon C, that is maintained constantly
in contact with such block.

EXTRACTS FROM STANDARD WORKS.

In a work entitled "Electric Light,"* by John W. Urquhart, the following description is given of an arc lamp on page 206 :

When two pointed sticks of carbon attached to the two poles of a source of electricity, such as any of those previously described, are touched together, a current will pass, and the carbons may then be separated a certain distance without interrupting the current, which is carried on by the intermediate air heated by the current, and an exceedingly brilliant light, which is termed the *voltaic arc*, will be produced between the carbons.

Particles of burning carbon are projected from one carbon to the other and a portion of the light is attributed to this flow of burning matter, but the greater portion is due to the incandescence of the carbon, or to a conversion of electric current into light, as inexplicable as that produced in a spark discharged between two conductors, or in a flash of lightning. The researches of Captain Abney, R. E., F. R. S., have shown that while the white light of the positive pole is always of the same composition in respect of the relative proportions of waves of different colours, the temperature of the arc from the graphite carbon is also the

---

*"Electric Light, Its Production and Use," by John W. Urquhart. London: Crosby Lockwood & Son. 1891. 407 pages, 145 illustrations. Price $3 00.

same in arcs of different powers—the temperature of fusing graphite.

The positive carbon, or that *from* which the current is generally assumed to flow, is, in voltaic arc lamps, consumed very fast, and becomes hollowed out, forming a crater, while the negative or receiving carbon is acted upon very slightly, and becomes pointed.   Carbon rods may burn at the rate of about five inches per hour, according to their size, and as they consume away must be fed up to each other in order to continue the light.   This was formerly done by hand, but now it is effected by such perfect automatic lamps that the light is not only perfectly steady, but needs no attention whatever for several hours together.   It is no difficult matter to feed carbons by hand, by means of a screw attached to one of the pencils, and for taking photographs by quick acting plates this will answer very well, but a lamp is the only satisfactory means by which ordinary carbon rods can be burned for general purposes.

VIII.—INCANDESCENT ELECTRIC LIGHTING.

In the incandescent electric lamp a strip or filament of carbon, or other refractory material, is heated to incandescence by the passage through it of an electric current.

In order to prevent the filament, when of carbon, from consuming or burning in the air, it is placed inside a lamp chamber from which all the air has been exhausted.

A substance suitable for use as the incandescing conductor of an electric lamp should possess the following properties:

(1.) It should be of high refractory power ; that is, it should be capable of being raised to a very high temperature without fusing or volatilizing.

(2.) It should possess a comparatively high electric resistance per unit of length.

(3.) It should be electrically homogeneous throughout all parts of its length.

(4.) It should be capable of being readily cut or fashioned into the required shape before carbonization.

Of the various substances that have been used for the conductors of incandescent lamps, carbon, obtained by carbonizing fibrous vegetable material, appears to most closely meet the above requirements.

Before being carbonized the fibrous material is cut into the required shape and then converted into carbon by any suitable carbonizing process.

Various shapes are given to the incandescent filament of the lamp. Some form of arc or horseshoe shape, however, is generally employed, so as to avoid the shadows which would otherwise be formed by the wires which carry the current into and out of the lamp chamber. In other words, an arc shape, or some modification of such shape, is necessary in order to permit the current to enter and leave the lamp chamber at points near together, so as to best avoid shadows, and make it convenient to secure the lamp to a socket, which holds it and contains the contacts for the circuit.

An incandescent electric lamp consists of the following parts, namely :

(1.) Of an incandescing conductor or filament through which the current passes.

(2.) Of an enclosed transparent chamber called the lamp chamber.

(3.) Of wires or conductors which pass through

the lamp chamber and are connected to the ends of the incandescing filament. These are called the leading-in wires.

(4.) Of various devices for supporting the filament inside the lamp chamber at its points of connection with the leading-in wires.

(5.) Of the lamp base, consisting generally of metallic points or rings cemented to the base of the

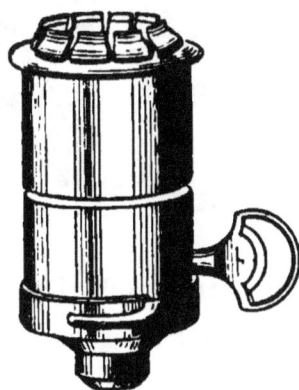

FIG. 78.—LAMP SOCKET.

lamp chamber and connected to the leading-in wires which pass through the lamp chamber.

(6.) Of the lamp socket, which consists of a device placed on the electrolier or bracket containing insulated contact plates or rings, connected to the terminals of the leads that furnish the lamp with current, so that, when the lamp base is merely placed in the socket, the leading-in wires of the lamp are connected with the leads.

Two well-known forms of lamp sockets are shown in Figs. 78 and 79.   Key switches are provided

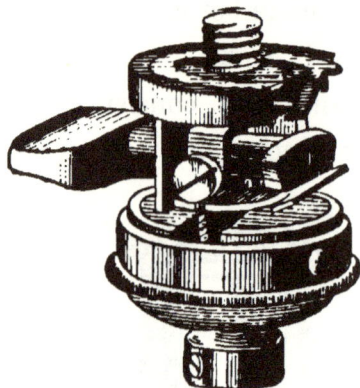

FIG. 79.—LAMP SOCKET.

for the purpose of turning off the current without removing the lamp from the socket.

The lamp sockets are supported on brackets, pendants or electroliers.   Some forms of lamp

FIG. 80.—LAMP BRACKETS.

brackets are shown in Figs. 80 and 81.   As in gas fixtures, the arms are either fixed or movable.

As generally employed the incandescing filament is produced as follows : Carefully selected, fine-grained bamboo is cut into strips of suitable length. All the softer material is removed from the inside of the strips as well as the hard siliceous outer coating. In this way a hard, fine-grained, homogeneous inner coating of fibrous material is obtained, which is then cut by means of planes or specially devised cutters into strips of the desired dimensions. In some forms of lamps the ends are made of greater

Fig. 81.—Lamp Bracket, Movable Arms.

dimensions in order to insure good electrical connection with the leading-in wires.

This latter point is of great importance in the proper operation of the lamp. As the current passes through the filament, it is necessary that the ends of the filament, where they are placed in connection with the leading-in wires, should not acquire too high a temperature, since otherwise the platinum of which such wires are formed would be liable to fusion. The increase of temperature at this point is avoided by making the area of cross section of the filament of such junction larger, thus rendering its conducting power greater.

The bamboo filament being suitably shaped is now subjected to a carbonizing process, which is carried on substantially as follows: The filaments being suitably shaped are bound or secured to the outside of a piece of carbon of the shape it is desired they shall have, and are then placed in boxes, covered with powdered plumbago or lamp-black, and subjected to the prolonged action of intense heat while out of contact with air. In this manner the filaments maintain their shape during carbonization.

The electrical conducting power of the carbon, which results from the carbonizing process, is increased by the action of heat, and also, in all probability, by the deposit throughout the mass, of carbon resulting from the decomposition of the hydro-carbon gases which are produced during the carbonization.

The carbon filaments so obtained are not yet suitable for use in the lamp. No matter how much care has been taken to select bamboo of uniform density, or in cutting it into strips of uniform area of cross section, it will be found, when such strips or filaments are raised to electric incandescence by the passage of a current through them, that they do not glow with equal brilliancy throughout all parts of their length, but certain portions are much brighter than others. If, for the purpose of rendering such conductors luminous throughout their entire length, the strength of the current is increased, the portions of high resistance would either fuse or volatilize, and the filament would be destroyed; or, if such portions only were allowed to give light, the lamp would not be economical in its working.

In order to avoid this difficulty and to render the conductor fit for actual use in the lamp, the filament is subjected to a process known technically as the flashing process.

In the flashing process the carbon filament is placed in a vessel filled with the vapor of a readily decomposable hydro-carbon, such as rhigolene, and gradually raised to electric incandescence by the passage of a current. A decomposition of the hydro-carbon occurs, the carbon resulting therefrom being deposited both in and on the conductor.

Ordinary incandescence, as by heat, would not answer for this process, since the heating is not properly directed. With electric incandescence, however, as the current is gradually increased the parts of the conductor where the electric resistance is the highest are first rendered incandescent and receive the deposit of carbon. As the current gradually increases, other portions become successively incandescent and receive a deposit of carbon, until at last the filament glows with a uniform brilliancy, indicative of its electric homogeneity.

After the flashing process the filaments are connected to the leading-in wires and placed in position in the lamp chamber. In order to insure a good electrical connection between the ends of the carbon filament and the leading-in conductors various devices are employed. One of these consists in inserting the ends of the carbon filament in cavities or spaces provided in the conductors, and depositing a layer of copper over the ends of the wire and the lower ends of the carbon filament by the process of electro-plating.

Another process consists in depositing a coating of carbon on the joint by immersing the filament at such point in a readily decomposable hydro-carbon liquid, as, for example, rhigolene, and passing a sufficiently powerful current through the joint to raise it to electrical incandescence. The carbon resulting from the decomposition is then deposited in a firmly adherent condition around the joint.

The leading-in wires are made of platinum, and are hermetically sealed in the lamp chamber by being fused to the portions of the glass through which they pass. Platinum is employed for this purpose because its rate of expansion is so nearly the same as that of the glass that it does not injure the vacuum in the lamp chamber when it alternately expands · and contracts on changes of temperature.

The mounted carbon filament being placed inside the lamp chamber the chamber is exhausted, first, by the action of a mechanical pump, by which the greater part of the air is rapidly removed, and afterward by the action of some form of mercury pump.

The form of pump frequently employed for this latter purpose is known as the Sprengel mercury pump.

In this pump, as shown in Fig. 82, the fall of a mercury stream causes the exhaustion of a reservoir connected with the vertical tube, at the point x, by the mechanical action of the falling mercury in entangling bubbles of air. The flow of the mercury can be started or stopped by means of a clip stop-cock. The bubbles of air are largest at the beginning of the exhaustion, and become smaller and smaller near the end, until at last the characteristic metallic click of mercury or other liquid falling in a good vacuum is heard. The exhaustion may be considered as completed when the bubbles entirely disappear from the column.

In actual practice the mercury that has fallen through the tube is again raised to the reservoir *A*, connected to the drop tube, by the action of a mechanical pump.

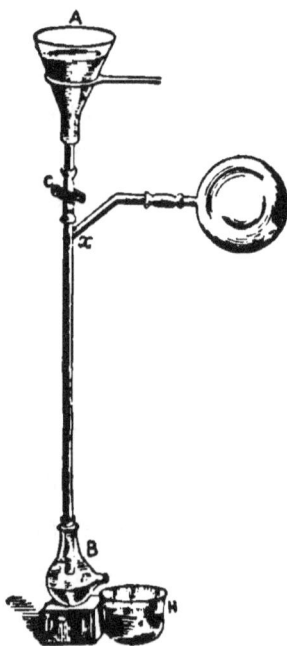

FIG. 82.—SPRENGEL'S MERCURIAL AIR PUMP.

Care must be taken to thoroughly remove all air from the chamber of the lamp. Since the filament is formed of carbon, if even a small quantity is left in the chamber the life of the lamp, or the time during which it can continue to act as an efficient source of light, will be greatly decreased.

Carbon possesses a marked power of taking in, absorbing or occluding gases, which it condenses in its pores. Even though the lamp chamber is exhausted of all the air it contains, if it is then hermetically sealed by the fusing of the glass, as soon as the carbon is raised to incandescence by the passage of the current through it, the occluded gas is driven out from the filament into the lamp chamber, and soon destroys the filament. In order to avoid this the lamp is subjected to an operation known as the occluded gas process.

This process consists essentially in raising the filament to incandescence by passing an electric current through it while the lamp is being exhausted, since otherwise the expelled gases would be re-absorbed. By this means a considerable quantity of occluded gas is driven out of the carbon which it would be impossible to get rid of by the mere operation of pumping.

Both the exhaustion and the incandescence continue up to the moment the lamp chamber is hermetically sealed ; otherwise some of the air might remain in the lamp chamber.

The lamp chamber is now hermetically sealed,

usually by the fusion of the glass in the manner adopted in the sealing of Geissler tubes or Crookes' radiometers.

Various forms are given to the incandescent filament. In the well-known form shown in Fig. 83 the filament has a horseshoe shape.

FIG. 83.—INCANDESCENT ELECTRIC LAMP.

In the Swan lamp, shown in Fig. 84, the filament is made of cotton thread in the form of a circular loop. This filament is made as follows: Cotton threads are immersed for a few moments in a mixture consisting of two parts of sulphuric acid and one of water, which converts the cellulose of the thread into artificial parchment. As soon as the filaments are removed from the sulphuric acid they are rapidly washed until all traces of the acid are removed. They are then passed through dies so as to insure a uniform area of cross section, and are wound

on rods of carbon or earthenware of the required outline, packed in a crucible filled with powdered charcoal, so as to exclude the air, and carbonized.

Incandescent lamps are generally connected to the
leads or circuits, either in multiple-arc or in multi-

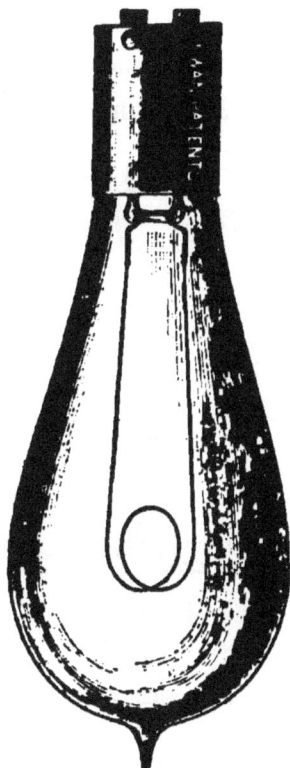

FIG. 84.—SWAN INCANDESCENT LAMP.

ple-series. They are, however, sometimes connected
to the line in series.

In practice it is usual to mark on incandescent

electric lamps the potential difference in volts which should be applied at the terminals in order to furnish the current necessary to properly operate them. If this potential difference is increased, the light emitted increases, but the life of the lamp is shortened.

When incandescent lamps are connected to the leads in multiple-arc, or in multiple-series, which are

FIG. 85.—SERIES INCANDESCENT ELECTRIC LAMP.

the connections generally adopted, the resistance of the filament is generally made high. When they are connected to such line in series, the resistance is generally made low. The resistance of the filament of a series-connected lamp is made low because the cross sections of the filament must be large to carry such large currents as are generally employed, and also because it is more convenient in practice to make such filaments short for the candle powers generally produced. A form of series-connected lamp is shown in Fig. 85.

In the case of the series circuit, when any receptive device is cut out or removed from the circuit, in order to prevent the opening or breaking of the rest of the circuit, a path must be provided by which the current can flow past the device thus removed or cut out. This is usually accomplished by means of a switch.

In a series-connected lamp some form of automatic switch or cut-out must be provided, which on the failure of any lamp in the circuit to properly operate will automatically cut such defective lamp out of the circuit, and will at the same time provide a by-circuit or path by which the current can flow past the faulty lamp and feed the remaining lamps placed in the circuit.

This is usually accomplished by some form of film cut-out in which the circuit is completed past the faulty lamp by piercing a sheet of paper or mica placed in a break in the circuit between two pieces of solder. On the piercing of the film the terminals

` are fused together and a permanent short circuit is effected past the lamp.

In a multiple circuit the opening of any lamp does not affect the rest which are still connected to the leads. In the multiple-series circuit the opening of a single lamp only affects the series circuit in which it is placed. Switches, therefore, are not required in such circuits.

The device employed for automatically breaking the circuit when the current has for some reason

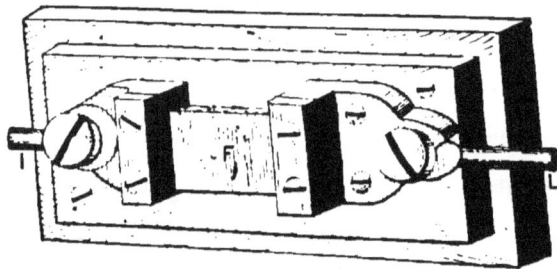

FIG. 86.—SAFETY FUSE.

become dangerously great consists of some form of safety fuse in which a strip, plate or bar of lead, or some other readily fusible alloy, that fuses and automatically breaks the circuit in which it is placed on the passage of an excessive current that would endanger the safety of other parts of the circuit.

Safety fuses are generally made of alloys of lead, and are placed in boxes lined with some non-combustible material in order to prevent fires from the molten metal.

Fig. 86 shows a fusible strip F, connected with leads L, L. Safety fuses are placed on all branch circuits and are made of sizes proportionate to the safe carrying capacity of the circuits which they guard.

The life of an incandescent electric lamp is reckoned by the number of hours during which it can furnish an efficient source of light. After being used for a time that will depend on the care with which the lamp has been constructed, the filament either breaks, or the lamp becomes useless from the chamber gradually becoming opaque.

The decrease in the transparency of the lamp chamber and the consequent decrease in the efficiency of the lamp may result either

(1.) From the settling of dust or dirt on the outer walls of the chamber; or,

(2.) From the deposit of metal or carbon on the inner walls of the chamber.

To obviate the first cause of diminished transparency, the outside of the lamp chamber should be frequently cleaned. The diminished transparency due to the second cause cannot be removed in the lamps in commercial use, and when it has reached a certain point, it is more economical to replace the old lamp by a new one.

In a properly made lamp the dimming of the lamp chamber is not apt to occur unless a stronger current than the normal current is passed through the lamp.

The life of an incandescent lamp should not be taken as the time which elapses until the filament actually breaks. As soon as the lamp chamber has become covered with such a deposit of carbon or coating of metal as to considerably decrease the

FIG. 87.—PORCELAIN LAMP SHADE AND WIRE GUARD.

amount of light which passes through the chamber, the lamp should be considered as useless.

The surface of the lamp chamber is sometimes ground so as to scatter or diffuse the light. Reflectors are sometimes placed back of the lamp for the purpose of throwing the light in one general direction. Porcelain shades are often employed to throw the light downward, as shown in Fig. 87, in which is also shown a wire shade guard provided for protecting the shade in exposed situations.

EXTRACTS FROM STANDARD WORKS.

In the second edition of his " Dictionary of Electrical Words, Terms and Phrases,"* on page 274, the author thus describes the properties which should be possessed by a good artificial illuminant :

Illumination, Artificial—The employment of artificial sources of light.

A good artificial illuminant should possess the following properties, namely :

(1.) It should give a general or uniform illumination as distinguished from sharply marked regions of light and shadow.

To this end a number of small lights well distributed are preferable to a few large lights.

(2.) It should give a steady light, uniform in brilliancy, as distinguished from a flickering, unsteady light. Sudden changes in the intensity of a light injure the eyes and prevent distinct vision.

(3.) It should be economical, or not cost too much to produce.

(4.) It should be safe, or not likely to cause loss of life or property. To this intent it should, if possible, be inclosed in or surrounded by a lantern or chamber of some incombustible material, and should preferably be lighted at a distance.

---

* " A Dictionary of Electrical Words, Terms and Phrases," by Edwin J. Houston, A. M. Second edition. New York: The W, J. Johnston Co., Ltd. 1892. 562 pages, 570 illustrations. Price $5.00.

(5.) It should not give off noxious fumes or vapors when in use, nor should it unduly heat the air of the space it illumines.

(6.) It should be reliable, or not apt to be unexpectedly extinguished when once lighted.

The electric incandescent lamp is an excellent artificial illuminant.

(1.) It is capable of great subdivision, and can, therefore, produce a uniform illumination.

(2.) It is steady and free from sudden changes in its intensity.

(3.) It compares favorably in point of economy with coal oil or gas, provided its extent of use is sufficiently great.

(4.) It is safer than any known illuminant, since it can be entirely inclosed, and can be lighted from a distance or at the burner without the dangerous friction match.

The leads, however, must be carefully insulated and protected by safety fuses. (See Fuse, Safety.)

(5.) It gives off no gases, and produces far less heat than a gas-burner of the same candle-power.

It perplexes many people to understand why the incandescent electric light should not heat the air of a room as much as a gas light, since it is quite as hot as the gas light. It must be remembered, however, that a gas-burner, when lighted, not only permits the same quantity of gas to enter the room which would enter it if the gas were simply turned on and not lighted, but that this bulk of gas is still given off, and is, indeed, considerably increased by the combination of the illuminating gas with the oxygen of the atmosphere; and, moreover, this great bulk of gas escapes as highly heated gases. Such gases are entirely absent in the incandescent electric light, and consequently its power of heating the surrounding air is much less than that of gas lights.

(6.) It is quite reliable, and will continue to burn as long as the current is supplied to it.

Slingo and Brooker, in a book entitled " Electrical Engineering,"* on page 544, speaks thus of the incandescent lamp :

Although it is, evidently, a comparatively simple matter to obtain the degree of exhaustion necessary for incandescent lamps, there are several causes for a deterioration manifesting itself in the vacuum after the finished lamp has been laid aside for a time, such as the occlusion of gases by the carbon and platinum, and by the cement employed to connect them together, and the very thin film of air which is liable to adhere to the inner surface of the bulb. In order to expel these gases, the filament is raised to incandescence during the later stages in the process of exhaustion, or the heat is applied externally.

The lamp having been sufficiently exhausted, the small glass tube connecting the bulb to the exhaust tube is fused, drawn out to a thread, and the lamp sealed off.

It remains now to test its efficiency, that is to say, the amount of light emitted for a given electrical power. A lamp may be said to have a very good efficiency if it yields

---

* "Electrical Engineering for Electric Light Artisan, and Students," by W. Slingo and A. Brooker. London: Longmans, Green and Co. 1890. 631 pages, 307 illustrations. Price, $3.50.

one candle power in return for 3.5 watts, so that an average 16 candle-power lamp should absorb 56 watts.

The vacuum is usually tested by means of an induction coil; one method is to fuse two platinum wires into a glass tube leading into the lamp, and simultaneously exhausted with it, and to connect these wires to the terminals of the secondary coils. The distance between the ends of the platinum wires inside the tube is so adjusted that when the required degree of exhaustion is attained, the spark passes through the air outside the bulb, in preference to traversing the vacuous space between the platinum points. Another method applicable to the finished lamp is to connect one end of the secondary to the filament. and the other to a loop wound outside the bulb, the quality of the vacuum being determined by the relative feebleness of the discharge which takes place between the filament and the bulb. It should be observed that, in a badly exhausted lamp, not only does the filament "burn," that is, oxidize, but it also requires a greater amount of heat to raise and maintain its temperature at the required point, owing to the fact that the air particles carry a portion of the heat away by convection.

IX.—ALTERNATING CURRENTS.

An alternating current of electricity is a current which flows alternately in opposite directions, as distinguished from a direct current which flows continually in one and the same direction ; in other words, an alternating current is a current whose direction of flow is continually reversed, such reversals rapidly and regularly following one another.

Since the current produced in the armature of a bi-polar dynamo-electric machine flows in one direction during its rotation past one of the poles of the field magnets, and in the opposite direction during its rotation past the other pole, the uncommuted currents from such machines are alternating or rapidly reversed currents.

In alternating currents the electromotive forces producing the current are directed in alternately opposite directions. In Fig. 88 these electromotive forces are represented in the form of a curve, the positive electromotive forces, or those which tend to produce a current in one direction, being represented by values above the line A E, and the negative electromotive forces, or those which tend to produce currents in the opposite direction, being represented by values below the line *A E*. The curves *A B C* and *C D E*, thus produced, are called the phases of the current *A B C*, the one above the line being generally called the positive phase, and *C D E*, the one below the line, the negative phase.

The phase, which is the time required to complete the to-and-fro motions above and below the

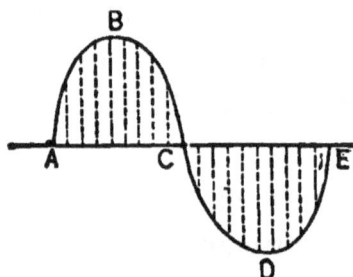

FIG. 88.—CURVE OF ELECTROMOTIVE FORCES OF ALTERNATING CURRENTS.

line *A E*, is practically the periodicity of the machine, and is dependent on the armature speed and the number of poles employed in the machine.

Two alternating currents are said to possess the same phase when the electromotive forces that produce them are simultaneously directed in the same direction.

Two alternating currents are said to possess the same period when the times during which the electromotive forces tend to produce currents in the same direction are equal. Two alternating currents are said to be in synchronism with each other when their electromotive forces tend to produce currents in the same direction and for the same length of time.

If two alternating dynamos be connected to the same leads in series, the electromotive force produced in such leads will theoretically be equal to the sum of the electromotive forces of the two machines. This, however, is only true if the phases of the two machines remain exactly the same. If they differ in even the slightest degree, they will rapidly tend to increase this difference of phase until they are in exactly opposite phases, when, of course, they will produce no current. Series connection or running of alternators is, therefore, impracticable.

If, however, two alternators be connected to the same leads in parallel, then, provided the armature circuits are so arranged that the currents in them can be rapidly reversed, and very small electromotive forces impressed on such circuits can produce large currents in them, and the engines driving the dynamos are under the control of the dynamos, that is, are not governed, then such machines, even if out of synchronism when coupled to the leads, will, almost immediately, pull each other into parallelism, each promptly exercising an automatic synchronizing control over the other.

The marked advantages possessed by alternating currents for certain kinds of electrical work have led to an extended study of their peculiarities. Such study has disclosed the fact that alternating currents differ in marked respects from the direct or continuous electric currents that have heretofore been almost exclusively employed in practical electrical work.

The following peculiarities concerning alternating currents should be carefully remembered :

(1.) The direction of the current undergoes regular changes.

(2.) The strength of the current undergoes regular reversals.

(3.) The peculiarities of the changes either in the direction of the current or in its strength during one complete alternation, or one complete to-and-fro motion, are regularly repeated during any subsequent to-and-fro motion.

A motion that regularly recurs or reproduces itself at regular intervals according to a certain law is called, in scientific language, a simple harmonic motion, or a simple periodic motion.

A pendulum that is set swinging in a circular path

affords an example of a simple-harmonic motion. If such a pendulum be looked at either from above or below, its path will appear to be circular. If looked at, however, from one side, its path will appear to be elliptical, and such elliptical path will appear longer and narrower as the eye of the observer approaches the level of the plane in which the bob of the pendulum moves, and, when it reaches this point, the bob will appear to move backward and forward in a straight line.

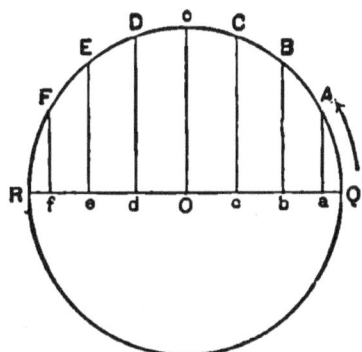

FIG. 89.—SIMPLE-HARMONIC MOTION.

Let the circle $Q o R$, Fig. 89, represent the path in which the bob moves, and let $Q A$, $A B$, $B C$, $C o$, etc., be equal distances in such path. Let the lines $A a$, $B b$, $C c$, $o O$, etc., be drawn perpendicular to the line $Q R$. Then, when looked at with the eye in the plane in which the bob travels, the line $Q R$, will be the path in which the bob appears to move backward and forward, and the lines $Q\,a$, $a\,b$, $b\,c$, $c\,O$, etc., will represent the spaces apparently traversed in equal intervals of time.

The circle $Q\,o\,R$, is called the circle of reference.

By a simple-harmonic motion is meant the motion which would result from the projection of the circular motion of the pendulum on the diameter $Q\,O\,R$, or the motion that is executed backward and forward along the line $Q\,O\,R$.

Examples of approximately simple-periodic or simple-harmonic motion are seen in the movements of the connecting rod of an ordinary steam engine or in the motion of the piston-rod of the steam engine.

The regularity with which both the variations in the strength of the alternating current and the changes in its direction recur, render the motions of such currents simple-harmonic or simple-periodic motions. Alternating currents are sometimes called simple-harmonic or simple-periodic currents. In many cases, however, the motions of alternating currents partake of the nature of complex-harmonic motions.

In the case of a direct or continuous current, the current strength is equal to the electromotive force divided by the resistance. In the case of an alternating current, since the electromotive force is constantly undergoing a change both in value and direction, to determine the current strength produced the average electromotive force must be divided by a quantity, allied to resistance, called impedance.

It is well known in the phenomena of electrodynamic induction, that when the current strength in any circuit undergoes variations the expanding and contracting lines of magnetic force cut portions of the circuit and produce electromotive forces, that tend to produce a current in one direction when such lines of force are moving inward or toward the conductor, and in the opposite direction when they are moving outward or from the conductor. Now this action of the current in inducing a current on itself, while its strength or duration is changing, which is called self-induction, or more simply inductance, exists in a marked degree in the alternating current. In any circuit, whatever be the kind of the current flowing through it, the passage of the current is resisted or opposed by the resistance of the conductor. In the case of an alternating current, besides this resistance there exists another apparent resistance which opposes the passage of the current ; namely, the self-induction or inductance produced by an electromotive force acting in such a direction as to oppose the passage of the current.   This is sometimes called the spurious resistance.

Impedance in any circuit may be defined generally as opposition to current flow.   The impedance is equal to the sum of the inductance and the ohmic resistance arising from the dimensions and character of the conductor.   In the case of any direct or continuous current, C, the current strength equals E, the electromotive force, divided by R, the resistance ; or,

$$C = \frac{E}{R}$$

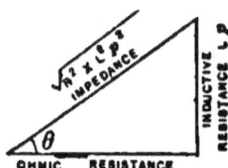

FIG. 90.—GEOMETRICAL REPRESENTATION OF IMPEDANCE.

In a simple-periodic or alternating current the average current strength

$$= \frac{\text{the average impressed electromotive force}}{\text{impedance.}}$$

The impedance is a quantity equal to the square root of the sum of the squares of the inductive resistance of the circuit and the ohmic resistance.

The impedance of a circuit can be represented

geometrically, as shown in Fig. 90. If the base of the right-angled triangle represents the ohmic resistance, and the perpendicular height the inductive resistance, then the hypotenuse, which is equal to the square root of the sum of the squares of the base and the perpendicular, equals the impedance.

A rapidly alternating current produces a variety of phenomena during its passage through a conductor, which differ markedly from the passage of a direct or continuous current through the same conductor. When a steady current flows through a conductor, the current density is the same for all areas of cross section. With a rapidly alternating current, however, the current density is greater near the surface, and, when the rate of alternation is sufficiently great, the current is almost entirely absent from the central portions of the conductor.

It has been shown by Lord Rayleigh that when the rate of alternations is 1,050 per second, the resistance of a conductor 100 millimeters in length and 30 millimeters in diameter is 1.84 times its resistance to direct or continuous currents. He found that such increase of resistance was greater with conductors of great diameter than with those of small diameter.

A careful study of some of the peculiarities of

alternating currents has led to a radical change of opinion concerning the manner in which the current is believed to flow through conducting paths. The current is not supposed to flow through the conductor, but to be propagated through the ether or other di-electric which surrounds the conductor and lies outside it. The conductor merely acts as a sink or place where the energy of the current is rained down upon it.

The current, or, perhaps, more correctly speaking, that which results in the current, is regarded as beginning at the surface of the conductor and more or less slowly soaking through it toward the centre. If the current is direct or continuous it soon reaches the deepest layers of the conductor; but if it is rapidly alternating, before it can soak very far into the conductor toward its centre it is turned back toward its surface, and so becomes confined to layers which will become more and more superficial as the rapidity of the reversals increases.

The conception of a rapidly alternating current flowing through a conductor by starting at the surface and gradually soaking in toward the centre does not regard the electric energy as moving through the conductor after the manner of water flowing through pipes, but as actually being rained down on its surface from the space outside of it.

This conception concerning the flow of alternating currents through a conductor is not unlike our idea of the flow of heat through a conducting wire, as has been pointed out by Stephan. If, for example, a wire or conductor which has been uniformly heated throughout be suddenly carried into a space where the temperature is higher than itself, the heat energy will pass into such wire from the surface toward the interior, the outer portions of the wire first rising in temperature and afterward the inner portions. In other words, in the case of a wire whose area of cross section is circular the heat penetrates toward the centre in successive concentric layers; or, conversely, if such a wire be carried into a space whose temperature is lower than the temperature of the wire, the wire or conductor parts with its heat energy by a movement which takes place through the wire outward.

Now, when the ends of a cylindrical conductor are subjected to an alternating electromotive force, by connection with the terminals of an alternating current dynamo, the energy of the current is believed to be conveyed to such conductor, not by actual passage through its substance, but through the space outside the conductor, said energy being rained down on its surface from the exterior, and gradually penetrating said conductor toward its centre. It will be seen that, in reality, a solid cylindrical conductor, which is conveying rapidly alternating currents, may in reality be regarded as a hollow cylinder of the same dimensions as the solid conductor, the thickness of the material in which will become smaller and smaller as the rapidity of alternation increases.

This conception concerning the passage of currents through conductors was first suggested by Poynting, and is known as Poynting's law.

The discharges of a Leyden jar partake of the nature of very rapidly alternating discharges. When such discharges are passed through the primaries of specially devised induction coils, as had been done by Nikola Tesla and Elihu Thomson, they produce all the phenomena of rapidly alternating currents in secondaries placed near them. As the electromotive force of such discharges is extremely high, it has been found necessary in practice to insulate the primary and secondary coils from each other by oil.

The phenomena of the alternative path of a discharge taken from a Leyden jar also demonstrates its oscillatory or alternating character. Such phenomena are seen in the lateral discharges that occur through the air space that separates such conductors from neighboring conductors. If, for example, a Leyden jar is provided with discharge wires or conductors, as shown in Fig. 91, the discharge is simultaneously attended by a large spark at *B,* between the ends of two long open-circuited leads.

The oscillations produced by the discharge in *A,* produce a pulsating field which induces oscillatory discharges in the open-circuited leads *B.* The counter-electromotive force produced in a conductor, through which an oscillating discharge is passing, by

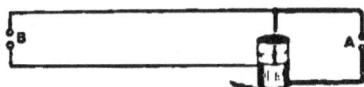

FIG. 91.—PHENOMENA OF ALTERNATIVE PATH.

the induction of the circuit on itself may render such conductor of much higher resistance than an intervening air space through which the discharge now takes place. These principles have been applied by Lodge and others to the construction and placing of lightning conductors so as to best enable them to act more efficiently in carrying the discharge from the neighboring cloud to the earth.

If the oscillatory discharge of a Leyden jar be

sent through a magnetizing coil, the magnetization
so produced on a steel rod placed inside such coil is
of so curious a nature that it was formerly described
as anomalous magnetization.   In such cases the bar
is not magnetized in the manner it would be if the
current passing through the magnetizing coil were
direct or continuous, but exhibits zones of positive
and negative magnetization extending from the surface of the bar inward toward the centre.   The explanation of this apparent anomaly is readily understood when the alternating character of a Leyden
jar discharge is borne in mind.   In other words, it
is not the magnetization that is anomalous.   If
there be any anomaly it is rather to be found in the
alternating or oscillating of the d scharge of the
Leyden jar.

   In order to avoid the effects of induction produced
in conductors which lie near other conductors,
through which alternating currents of electricity of
very high frequencies are passing, it is only necessary
to place on the outside of such conductors a conducting coating or covering of metal insulated therefrom.   The thickness of said conductor will necessarily depend on its conducting power.   This thickness, however, will be smaller in proportion as the
rapidity of the alternations increases.   As a rule, lead.

coated conductors do not act efficiently for such purposes on account of the low conducting power of lead.

The manner in which a conducting plate of metal acts to protect the conductor from the effects of a neighboring conductor through which a rapidly alternating current is passing, when such a plate is placed between two conductors, is as follows : Before the expanding or contracting lines of force can cut or pass through the neighboring conductor, they expend their energy in producing currents by induction in the material of the interposed plate. If such plate is of sufficient thickness, all their energy is expended on the interposed plate, and the neighboring conductor is screened or protected from such action.

In other words, the screening is due to the production in the interposed plate of currents called eddy currents, as can be proved by the fact that when such screen is split radially from the circumference to the centre, its screening effect is removed, since then such eddy currents cannot be produced.

When the screening plate is formed of iron it produces an additional effect from the tendency of such a plate to condense the lines of magnetic force on it by reason of the small magnetic resistance which iron offers to their passage through it.

The magnetic screens used for watches consist essentially of iron cases, which protect the magnetized portions of the works by closing the lines of magnetic force through the iron case.

EXTRACTS FROM STANDARD WORKS.

The conditions which are now generally assumed to attend the transmission of an alternating current through a conductor are thus happily expressed by Fleming in his "The Alternate Current Transformer,"* Vol. I., on page 476:

The whole history of the discharge may be divided into three parts. First, a time when the energy associated with the system is nearly all electrostatic and is represented by the energy of the lines or tubes of electrostatic induction running from plate to plate ; second, a period when the discharge is at its maximum, when the energy exists partly as energy associated with lines of electrostatic induction expanding outward, and partly in the form of closed rings or tubes of magnetic force expanding and contracting back on the wire ; and then, lastly, a period when nearly all the energy has been absorbed or buried in the wire, and has there been dissipated in the form of heat, which is radiated out again as energy of dark or luminous radiation. The function of the discharging wire is to localize the place of dissipation and also to localize the place where the magnetic field shall be most intense, and all that observation is able to tell us about a conductor which is conveying that

---

* "The Alternate Current Transformer in Theory and Practice," by J. A. Fleming, M. A., D. Sc. 2 vols. London: The Electrician Printing and Publishing Co. 1889-1892. Vol. I., 487 pages, 157 illustrations. Price $3.00. Vol. II., 594 pages, 300 illustrations. Price $5.00.

which is called an electric current is that it is a place where heat is being generated and near which there is a magnetic field. These conceptions lead us to fresh views of very familiar phenomena. Suppose we are sending a current of electricity through a submarine cable by a battery, say, from zinc to earth, and suppose the sheath is everywhere at zero potential, then the wire will be everywhere at a higher potential than the sheath, and the level surface will pass through the insulating material to the points where they cut the wire. The energy which maintains the current, and which works the needle at the further end travels through the insulating material, the core serving as a means to allow the energy to get into motion or to be continually propagated. This energy sucked up by the core is, however, transformed into heat and radiated again as dark heat. If we adopt the electro-magnetic theory of light, it moves out again still as electro-magnetic energy, but in a different form, with a definite velocity and intermittent in type. We have, then, in the case of the electric light this curious result—that energy moves in upon the arc or filament from the surrounding medium, there to be converted into a form in which it is sent out again, and through which the same in kind is able to affect our senses.

In the case of an arc or glow-lamp worked by an alternating current, we have still further the result that the energy which moves in the carbon is returned again, with no other change than that of a shortened wave-length, and the carbon filament performs the same kind of change on the electro-magnetic radiation as is performed when we heat a bit of platinum foil to vivid incandescence in a focus of dark heat. A current through a seat of electromotive force is therefore a place of divergence of energy from the conducting circuit into the medium, and this energy travels away and is converged and transformed by the rest of the circuit. From this aspect the function of the copper conducting wire fades into insignificance in interest in comparison with the function of the dielectric, or rather of the ether contained in the dielectric. When we see an electric tramcar, or motor, or lamp worked from a distant dynamo, these notions invite us to consider the whole of that energy, even if it be thousands of horse-power per hour, as conveyed through the ether or magnetic medium, and the conductor as a kind of exhaust valve, which permits energy to be continually supplied to the dielectric.

In his " Dictionary of Electrical Words, Terms and Phrases,"* on page 467, the author thus defines and explains magnetic screening:

Screening, Magnetic.—Preventing magnetic induction from taking place by interposing a metallic plate, or a closed circuit of insulated wire, between the body producing the magnetic field and the body to be magnetically screened.

A magnetic needle is screened from the action of the earth's field by placing it inside a hollow iron box, which

---

prevents the lines of force of the earth's field from passing through it by concentrating them on itself. This action is dependent on the fact that iron is paramagnetic and therefore offers the lines of force less resistance through its mass than elsewhere. A plate of copper would not effect any such magnetic shielding or screening.,

In any magnetic field, however, in which the strength of the field is undergoing rapid periodic variations, a plate of copper or other electric conductor may act as a screen to protect neighboring conductors from the effects of magnetic

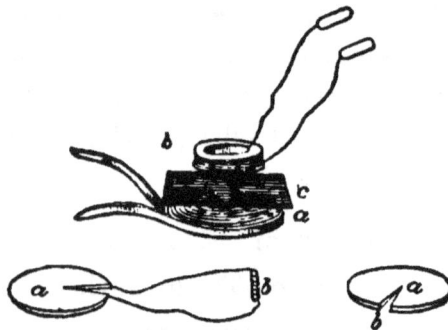

FIG. 92.

induction, and its ability to thoroughly effect such a screening will depend directly on its conducting power.

If, for example, the copper plate c, (Fig. 92) be interposed between a coil of copper ribbon a, and the fine wire coil b, it will greatly reduce the intensity of the induced currents produced when rapidly alternating currents are sent through a. If, however, the copper plate be slit, as shown to the right at a, the screening effect is lost, but is regained if the slit be connected by a conductor. Similarly a flat coil of insulated wire effects no screening action when open, but when closed acts as the uncut copper plate.

Here the screening action is due to the fact that the energy of the field is spent in producing eddy currents in the interposed metal screen or coils. If the metal screen is discontinuous in the direction in which the eddy currents tend to flow, the inability of the screen to absorb the energy as eddy currents prevents its action as a screen.

The term magnetic screening is generally employed in the latter sense of preventing magnetic induction from occurring in a neighboring conductor, by interposing some conducting substance in which eddy currents can be freely established.

As to the efficiency of the screening action, if the makes-and-breaks do not follow one another very rapidly, the following principles can be proved :

(1.) If the screening material have absolutely no electrical resistance it will effect a perfect magnetic screening when placed between the primary and secondary, no matter what its thickness may be.

(2.) If the screen have a finite conductivity the screening will be imperfect, unless the thickness of the material employed is considerable.

If, however, the makes-and-breaks follow one another very rapidly, then

The screening effect of even imperfect conductors will become manifest with comparatively thin screens of metal.

As to magnetic screening, therefore, it follows that the less the conductivity the greater must be the speed of reversal, in order that the screening action may be effective.

Where a screen of iron is employed, an additional effect is produced by the fact that the small magnetic resistance of the metal, or its conductivity for lines of magnetic force, causes the lines of induction to pass through its mass, and thus effect a screening action for the space on the other side. This action is, by some, called magnetic screening.

In the case of iron screens, considerable thickness is required in the metal plate, in order to obtain efficient screening action of this latter character. On account of this action of iron, in conducting away lines of force, a much smaller speed of reversal is required in order to obtain effective screening action where plates of iron are used than in the case of plates of other metal.

FIG. 93.—WILLOUGHBY SMITH'S APPARATUS.

The apparatus shown in Fig. 93 was employed by Mr. Willoughby Smith in studying the effects of magnetic screening.

The flat coils A and B, were employed for the primary and secondary coils respectively, and were connected to the battery C, and the galvanometer F, as shown. Current reversers D and E, were so arranged as to reverse galvanometer and battery alternately, and so cause the opposite induced currents to affect the galvanometer in the same direction.

X.—ALTERNATING CURRENT DISTRIBU-TION.

Some of the peculiarities of alternating currents render their employment for the distribution of electric energy over extended areas very advantageous in the case of many of the practical applications of electricity.

A system for the distribution of electricity by means of alternating currents embraces the following parts, namely :

(1.) An alternating current dynamo-electric machine, or battery of machines.

(2). A pair of conductors or line wires arranged in a metallic circuit.

(3.) A number of transformers whose primary coils are placed in the circuit of the line wires.

(4.) A number of electro-receptive devices placed in the circuit of the secondary coils of the transformers.

(5.) Instruments for maintaining constant either the current or the potential, despite changes in the load on the consumption circuit.

There are two distinct systems of distribution by means of which the effects of alternating currents can be employed at points situated at fairly considerable distances from where the sources are located, namely :

(1.) A system of constant potential distribution.

Here the primary coils of a number of transformers are connected in multiple to leads which connect the distant stations where the source of alternating currents is located.

Such leads are maintained at an approximately constant average electromotive force, or potential. In constant potential distribution the primary circuits of the transformers consist of a great length of comparatively thin wire and the secondary coils of a comparatively short length of thick wire. This system of distribution is especially adapted for the operation of incandescent lamps or other receptive devices that are connected to the leads or conductors in multiple.

(2.) A system of constant current distribution.

Here the primary coils of a number of transformers are connected in series in a single metallic circuit. In this case the primary coils consist of short thick wires and the secondary of long thin wires. Such a system is especially adapted for the distribution of arc lights.

Let us now examine the advantages to be derived from the distribution of electricity by means of alternating currents employed in connection with leads maintained at approximately constant differences of potential, in cases where the distance between that part of the line where such currents are generated and that part at which they are to be utilized is fairly considerable.

During the passage of an electric current through any circuit, the energy which is expended in any particular parts of the circuit is in direct proportion to the resistance of such parts. Now, in any system of distribution economy of distribution requires that the energy expended in the electro-receptive devices shall bear as large a proportion as possible to the energy expended in the entire circuit. No difficulty exists in obtaining economical conditions in series-distribution circuits, even though their length is considerable, for the total resistance of such a circuit increases with every electro-receptive device added. Therefore, even though comparatively thin wires of great resistance are employed to connect such separate electro-receptive devices in series, yet the greater proportion of such resistance can still be placed in the devices themselves, and, therefore, a considerable extent of territory can be covered by such circuits without very great loss.

In systems of multiple distribution, however, an indefinite increase in the number of electro-receptive devices is impossible, because every electro-receptive device added decreases the total resistance of the circuit, and, unless the resistance of the leads is correspondingly decreased, the economy becomes smaller, unless, of course, the resistance of the leads was originally so low as to be inappreciable when compared with the change of resistance.

In systems of distribution by means of transformers, alternating currents, of small current strength and considerable difference of potential, are sent over a line from a distant station, and passing through the primary coils of a number of step-down transformers, generally connected to the line in multiple-arc, produce, by induction, currents of comparatively great strength and small difference of potential in the secondary coils. The electro-receptive devices are connected in multiple-arc to the circuits connected to the terminals of the secondary coils.

In cases where the distance is considerable, this method of distribution is employed and greatly reduces the cost of the main conducting wires or leads, because the current strength in the line wire is very small as compared with its value after conversion.

The general arrangement of the transformers on the main line, the secondary circuits and the connection of the electro-receptive devices to the coils of the secondary circuits are shown in Fig. 94. The transformers are supported on line poles, as

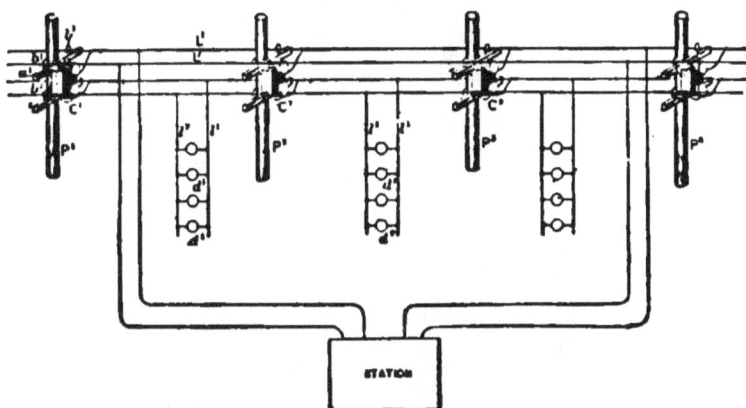

FIG. 94.—DETAILS OF TRANSFORMER CIRCUITS.

more clearly shown in Fig. 95, in which the terminals of the primary and secondary of the transformer are readily seen.

When the primary circuits of the transformers are connected in multiple-arc to leads or conductors that are maintained at approximately constant differences of potential, and the lamps or receptive devices are also connected to the leads in multiple, then, when the impedance of the primaries of each transformer is large enough, when its secondary is open, to keep practically all current from the primary, the amount of current which can flow through the primary circuit of such transformers is automatically limited to

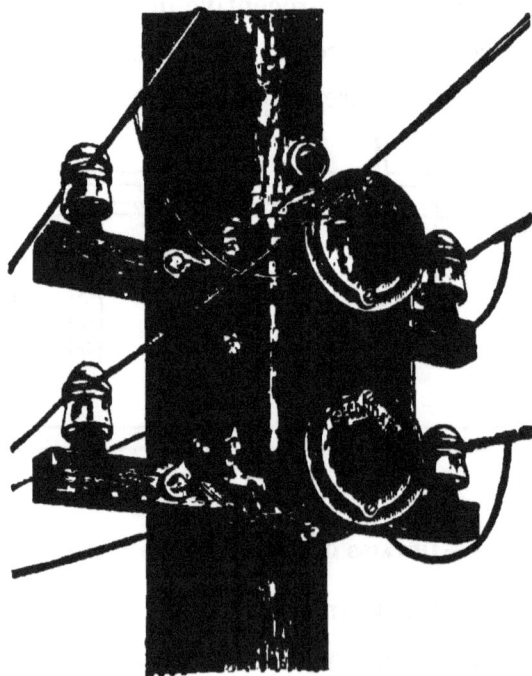

FIG. 95.—PRIMARY AND SECONDARY CIRCUITS OF TRANSFORMERS.

the actual requirements of the devices connected in multiple-arc to the secondary circuits of the transformers, and the system becomes automatically self-regulating.

If, for example, a full load is placed on such a line, and lamp after lamp is turned out, the current in the primary will decrease, a smaller current passing into the primary from the constant potential leads, and, at last, when all such lamps are turned off, although the primary circuit is still connected as before to the mains at full pressure, yet no current will flow through such primary, and, therefore, no current will be induced in the secondary.

The explanation of this curious automatic regulation is to be found in self induced counter-electromotive forces set up in the primary, which effectually bar the current from flowing through the mains.

When the lamps are entirely removed from the secondary circuit, the secondary circuit being open, the primary circuit can generate no current therein. The lines of force of the alternating primary circuit then expend their energy on the iron core placed inside the primary, which sets up counter-electromotive forces therein which act as a resistance to prevent any but a very small current from flowing through the primary from the leads.

There may be some difficulty in understanding why alternating currents should produce their maximum effect, as to the magnetization of the core, when the lamps are not in action. The reason is given by S. P. Thompson substantially as follows: The currents in the secondary are transformed into almost exactly opposite phases as those in the primary. While, therefore, the strength of the current in the primary is increasing, the strength of the current in the secondary circuit is increasing in the opposite direction; that is, the current flows through the secondary circuit in practically the opposite direction to what it does through the primary; and, of course, the magnetizing effect on the core is less. Therefore, as the load of lamps increases the magnetizing effect on the core decreases, and the counter-electromotive forces induced in the primary decrease and thus permit a stronger current to flow therein.

There occurs, however, a drop of potential on the mains conveying alternating currents which increases with the load. When the variations in the load are great the drop is sufficiently great to need an increase of potential at the generating end of the line. This increase is obtained either by an increase in the charge of the field or by the use of a regulator. In the Stillwell regulator, a transformer, provided with a secondary of variable length, is so connected with the line as to add its electromotive force thereto. The required length of secondary is introduced into the circuit by means of a lever switch. When such a regulator is used a compensator is required for the station voltmeter.

The electricity required for use in systems of distribution by alternating currents is generally obtained by means of an alternating current dynamo-electric machine, or by a battery of such machines. The rapidity of alternation generally employed for such purposes is comparatively great. To readily obtain such rapidity of alternation it is necessary that the armature rotate very rapidly in the magnetic field of the field magnets so as to increase the number of times it alternately passes north and south poles in such field.

The rate of alternation required in practice is so great that with a single pair of poles in the field of the dynamo the peripheral speed of the armature would be inconveniently great. In order to avoid this some form of multipolar machine is used.

Various forms are given to alternating current dynamo-electric machines. Any form of bi-polar or multi-polar dynamo-electric machine is capable of producing alternating currents provided its field magnet coils are suitably excited. It is only necessary to connect all the positive ends of the armature coils in a certain position of rotation to a conducting ring fixed to the shaft of the machine, and all the negative ends of the coils to another similar ring, and to carry off the alternating currents therein produced by means of collecting brushes resting on such rings. These rings will then become alternately positive and negative on the rotation of the armature, and will, therefore, furnish alternating currents to the brushes resting on them.

It has been found in practice, however, most convenient to devise special forms of dynamo-electric machines for the production of alternating currents. In a well-known form of alternating current dynamo-electric machine, the field magnet frames contain a series of magnet poles which project radially inward from the inside of the frame. These poles are alternately of north and south polarity.

The armature is formed of a drum composed of discs separated by some insulating material, and provided with holes for ventilation. The armature coils are formed of heavy conductors laid on the surface of a drum, to which they are securely attached, in order to prevent their separation during rotation.

The number of armature coils is equal to the number of field magnet poles. In some cases an alternating current dynamo is separately excited; in other cases some of the armature coils are suitably connected to commutators so as to furnish the direct currents that are employed for the magnetization of the field magnets.

The current produced in the separate coils of the armature may, before connection to the collection ring, be either connected with one another in series

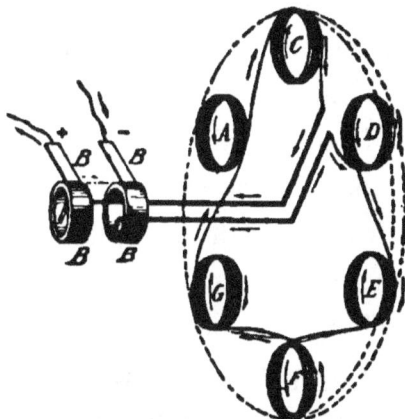

Fig. 96.—Circuit Connections of Armature Coils.

or they may be connected in multiple arc. The manner of their connection in series will be understood from an inspection of Fig. 96.

Since the electromotive forces employed in the distribution of alternating currents by transformers connected in multiple to constant potential leads are considerable, care must be taken to avoid the entrance

of such high potential currents into the buildings
where the low potential currents produced in the
secondary are utilized.   For this purpose the transformers should either be placed on the outside of
the building or on poles, where they are out of reach
and are thus placed out of accidental contact or tam-
pering by unauthorized persons, and the secondary
circuits be thoroughly insulated from the primary.
Elihu Thomson places a metallic, earth-connected
plate between the primary and secondary circuits in
order to discharge the said primary current to the
earth in case of accidental contact.

During action a disagreeable musical note, the
pitch of which varies with the rapidity of the alternation, is often emitted by the transformers.   The
disagreeable effects of this note may be, to a great
extent, avoided by placing the transformers in non-
elastic connection with their supports.

When an alternating current is sent through an incandescent lamp, although its filament alternately
increases and decreases in temperature, and the light
it emits undergoes similar changes in intensity, yet
the effect which such light produces on the eye will
be that of a steady, non-flickering light, provided
the rate of alternation is sufficiently rapid.   A similar effect is produced in the Jablochkoff candle, in

which alternating currents are used. In order to insure this steadiness in the illuminating power, the alternations should not follow one another too slowly. A fairly steady light is obtained with as few as 80 reversals per second. A higher rate, however, is preferable.

In direct or continuous current distribution of electricity, by means of transformers, a number of different systems have been devised. In one of these systems devices called motor-generators are employed. In this system of distribution by motor-generators a continuous current of high potential is distributed through a main line or conductor, and at the points where this energy is to be consumed, the high tension current is utilized for driving a motor, which in turn drives a dynamo, the current of which is of low tension and great quantity, suitably employed for feeding the electro-receptive devices.

In some cases the motor and generators are combined in a single double-wound armature, the fine wire coils of which receive the high potential driving current, and the coarse wire coils furnish the low potential currents used in the distribution circuits.

In another system of distribution by means of continuous currents, such currents are sent over the line, and, at the distant point where it is desired to utilize their energy, a device called a disjunctor is employed to rapidly and periodically reverse their direction. The alternating currents so obtained are used either by means of suitable transformers to change the character of the electromotive force and the current strength to meet the requirements of the distribution circuit, or are employed directly to charge condensers in the circuits of which induction coils are placed. Sometimes, however, the opposite plates of the condensers are connected directly to the incandescent electric lamps.

When it is desired to obtain a greater current strength on an alternating current circuit than can be supplied by a single dynamo, it is necessary to couple or connect two such dynamos in multiple or parallel. To do this it is not necessary that the dynamos be first synchronized; for, if the armature circuits can have their currents rapidly reversed, and small electro-motive forces impressed thereon produce large currents and the driving engines are not governed, then the machines, even if out of synchronism when coupled, will rapidly pull each other into synchronism.

The peculiar effects of self-induction produced by alternating currents render it possible to prevent variations in the light emitted by a number of incandescent lamps placed in parallel on any single distribution circuit.

A device frequently employed for such purposes is called a choking, kicking, or reaction coil. Such coils are used to obstruct, choke or cut off an alternating current with a loss of power less than by the use of a mere ohmic resistance.

A choking coil is shown in Fig. 97. It consists of a circular solenoid of insulated wire, wound on a core of soft iron wire, the separate turns of which are

FIG. 97.—CHOKING-COIL.

insulated from one another. By using iron wire for the core no eddy currents are produced in the coil.

The higher the periodicity the greater is the choking effect of a given coil, or the smaller the coil may be made in order to insure a given effect.

The choking coil operates by generating a counter-electromotive force, which tends to balance the applied electromotive force, and therefore cuts it down or chokes it.

Since a magnetic circuit completed partly through

iron and partly through air requires a greater current to produce saturation than a closed magnetic circuit, the throttling or choking power of such a coil is increased by forming its core in a closed magnetic circuit ; that is, of a circuit in which there is no air-space or gap.

A choking coil is sometimes employed for the purpose of varying the intensity of the light emitted by incandescent lamps in a device known as the dimmer.

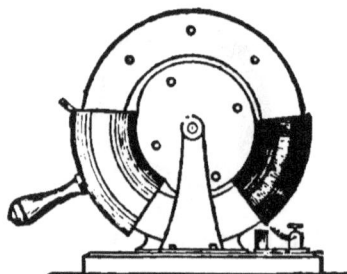

FIG. 98.—THE DIMMER.

The dimmer consists, as shown in Fig. 98, essentially of a choking coil wound around a portion of a laminated ring of soft iron. A laminated drum of iron is placed inside the ring, and suitably supported in bearings. There is fastened to the drum a heavy copper sheath, which is rotated with it. By moving this sheath so as to slide over and cover more or less of the coil, the self-induction in the coil becomes less, and therefore the current which can pass through it will become greater; when the sheath is moved away from the coil, the current which can pass becomes less. The dimmer is used in theatres or similar situations to turn the lights up or down, and in central stations for adjusting the difference of potential of the feeders.

The balanced reactive coil, an invention of Prof. Elihu Thomson, is a device for maintaining a constant current in the secondary, and is shown in Fig. 99. It

FIG. 99.—BALANCED REACTIVE COIL.

consists essentially of a choking coil, which is so counter-balanced as to automatically adjust the potential in a circuit of lamps.

In this form the coil is provided with a metallic sheath which is maintained in a balanced position by the action of a counter weight at *P*, and the spring *S*.

When a lamp is extinguished in the circuit the

variation in current due to such variation in resistance causes the sheath to be deflected and thus increases the self-induction of the coil and reduces the current in the lamp circuit to its normal value.

.

EXTRACTS FROM STANDARD WORKS.

The advantages to be derived from the use of high potential alternating currents in systems of distribution, when the electro-receptive devices are situated at comparatively great distances from the source, is thus discussed by Urquhart in his " Electric Light,"* on page 191.

In the practical distribution of electrical currents for lighting it was soon found that to convey large currents at low potential to a distance, conductors so large as to be impracticable were required. On the other hand, although it was well known that small currents of high tension representing the same amount of energy could be conveyed easily in exceedingly small conductors, the principle could not be applied to ordinary lamps direct, and the introduction of such currents into dwellings would be a possible source of danger to life.

It has long been known that a low tension current could by suitable means be converted into a high tension current.

The *induction coil*, an instrument for this purpose, is too well known to call for description. Its theory has been exhaustively treated in most text-books of electricity. The most powerful machine of this kind was owned by the late

*"Electric Light: Its Production and Use," by John W. Urquhart. London: Crosby Lockwood & Son. 1890. 380 pages, 145 illustrations. Price $3.00.

Mr. Spottiswoode, F. R. S. This coil would convert a low tension and harmless current into a high tension discharge, which would flash across an air-space 45 inches in width. Thus, from a current of a few volts a conversion was made to a current of many million volts.

But the induction coil is reversible, for, by feeding its secondary coil with a high tension alternating or interrupted current, a low tension current of great volume is obtainable from the primary coil.

Thus, the induction coil, which until within a few years ago was but a scientific toy, has developed into a most important auxiliary to the dynamo-electric machine.

In so far as the use of *transformers*, as they are generally called, has taken effect, they have only been successfully used for alternating currents. It is well known that if a constant current be passed through the primary wire of an induction coil the secondary circuit will evince no sign of current. At the moment of making or breaking contact with the primary coil, however, momentary currents will flow in the secondary coils. Hence the necessity to use a contact breaker or interrupter with such coils.

But if a constant current flows in the primary coil, and that coil be moved within the secondary coil, currents corresponding to the motions will be induced in the secondary —in fact, we have now a kind of dynamo machine. Hence, if the current transformer can be used for converting constant currents of high force to constant currents of low force, they must take the form of machines of some kind.

XI.—ELECTRIC CURRENTS OF HIGH FREQUENCY.

When, in the case of alternating currents, the rate of alternation and the electromotive force become very high, a number of phenomena present themselves that differ in many respects from those of alternating currents of only moderately high frequency. Some of these differences may be noticed as follows :

(1.) When alternating currents of very small frequency are sent through an electro-magnet, and a piece of iron is held against its poles, the alternate attractions and repulsions of its armature can be readily felt by the hand ; if the frequency of the alternations is increased, the impulses felt follow one another faster, but become weaker, and, when the frequency becomes much higher, there is apparently nothing but attraction, which manifests itself by a continuous pull.

(2.) Many substances which act as insulators for steady currents will permit alternating currents to readily pass through them, provided the frequency and difference of potential are sufficiently high. This is especially the case if a gas be present. The discharge will then take place by molecular bombardment through considerable thicknesses of such insulators as glass, hard rubber, sealing wax, etc.

Therefore, in the insulation of conductors conveying discharges of very high frequencies, where the differences of potential are great, all air and gas must be carefully excluded.

(3.) Conducting substances which readily permit the passage of steady currents offer a resistance to alternating currents which increases in amount as the frequency of the alternations increases, until at last, when the frequency of alternation has reached certain high limits, such substances act practically as non-conductors.

(4.) The physiological effects of alternating currents of but moderate frequency are unquestionably severe. When, however, the rapidity of alternation increases, alternating currents undoubtedly become less severe in their physiological action, and this decrease becomes the more marked as the frequency of alternation increases, until finally they apparently become absolutely harmless.

This is probably due to the fact that, when such increase of rapidity of alternation is sufficiently great, only the superficial portions of the body are affected, the increase in rapidity of alternation producing a counter-electromotive force or spurious resistance, which opposes the passage of the current through the body.

Nikola Tesla was the first to make extended investigations in the field of alternating currents of exceedingly high frequency, and it is to him that the credit is due for most of the above-mentioned facts concerning the difference in the behavior of alternating currents of exceedingly high and of but moderately high frequency.

Tesla obtained the exceedingly high frequencies with which he experimented in the following manner : By means of a multipolar dynamo-electric machine, the armature of which was driven at a high peripheral speed, and the currents of which were uncommuted, he obtained alternating currents of very high frequency. These currents were sent through the primary circuit of a peculiarly constructed induction coil. Under these circumstances Tesla noticed a number of exceedingly peculiar effects in the luminous character of the discharge taken between the secondaries of the induction coil.

Tesla describes five distinct characters of high frequency luminous discharge ; namely :

(1.) *The sensitive-thread discharge,* in which a thin, thread-like discharge occurs between the terminals of the secondary of the induction coil as soon as a certain high frequency is reached.

According to Tesla, this discharge occurs when the number of alternations in the primary is high, and the strength of the current passing is small. The discharges present the appearance of thin, feebly colored threads. Though easily deflected by

FIG. 100.—SENSITIVE-THREAD DISCHARGE (TESLA).

the breath, such discharge is quite persistent, resisting efforts to blow it out, provided the terminals are placed at one-third the striking distance apart. Tesla ascribes its extreme sensitiveness, when long, to the motion of dust particles suspended in the air through which the spark is passing. The general appearance of the sensitive-thread discharge is shown in Fig. 100.

(2.) *The flaming discharge,* which takes place when the current through the primary of the induction coil is increased. Under these circumstances the thickness of the discharge increases until it assumes a flaming white, arc-like discharge.

According to Tesla, the flaming discharge occurs when the number of alternations per second is great, and the other conditions of the circuit are such as will permit the passage through the primary of the coil of the maximum current.

FIG. 101.—FLAMING DISCHARGE (TESLA).

The flaming discharge develops considerable heat. The shrill note that accompanies less powerful discharges is absent in the flaming discharge. This, most probably, results from the enormous frequency of the discharge carrying the note far above the limits of audition.

Some idea of the general appearance of the flaming discharge may be had from an inspection of Fig.

101. According to Tesla, the conditions for its production, in an ordinary induction coil of say 10,000 ohms resistance, is best obtained by a rate of alternation of about 12,000 per second.

(3.) *The streaming discharge,* which occurs as the frequency of the discharge through the secondary increases and the current strength decreases. The streaming discharge partakes of the general nature of the flaming discharge. Luminous streams pass

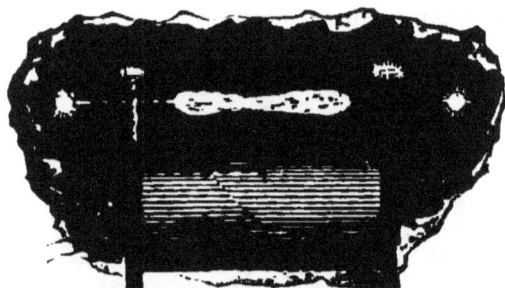

FIG. 102.—STREAMING DISCHARGE (TESLA).

in abundance, not only between the terminals of the secondary, but even between the primary and the secondary through the insulating dielectric substances separating them. They also issue from all points and projections, as shown in Fig. 102.

(4.) *The brush-and-spray discharge,* which occurs when the streaming discharge reaches a certain higher limit.

The streaming discharge that can be obtained

from an induction coil with high frequencies differs from that obtained from an electrostatic machine in that it neither possesses the violet color of the positive discharge nor the brightness of the negative discharge, but is paler in color.

The appearance of the brush-and-spray discharge is shown in Fig. 103.

The brush-and-spray discharge, when powerful, closely resembles a gas flame issuing from a gas

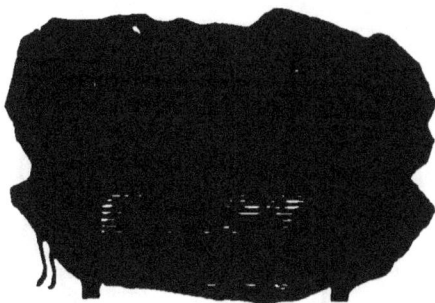

FIG. 103.—BRUSH-AND-SPRAY DISCHARGE (TESLA).

burner under great pressure. Speaking of such discharges Tesla says :

" But they do not only *resemble*, they *are* veritable flames, for they are hot. Certainly they are not as hot as a gas burner, *but they would be so if the frequency and the potential would be sufficiently high.*"

(5.) *Tesla's fifth form of discharge.* This form of discharge occurs as a change of the brush-and-spray discharge when the frequency of the discharge through the primary is still further increased. The terminals now refuse to give sparks except at very short distances, probably from the exaggerated tendency to dissipate. At this stage the discharge appears to pass through thick insulators with great readiness. The appearance of this form of discharge is shown in Fig. 104.

Beautiful luminous phenomena are produced by interposing or placing insulating substances between

FIG 104.—FIFTH TYPICAL FORM OF DISCHARGE (TESLA).

the terminals of a coil arranged so that this fifth form of discharge has been obtained. If, for example, a thin plate of ebonite is placed between the terminals, each of which has been provided with metallic spheres, as shown in Fig. 105, the sparks cease, and, provided the spheres are sufficiently large, the discharge produces instead an intensely luminous circle several inches in diameter.

Under certain circumstances the brush discharge

may be obtained in a very powerful form. This is best obtained when the terminals of the secondary are attached to bodies whose surfaces can be adjusted to give the best results with that particular coil and that particular discharge. In this case the brush discharge becomes very marked, and gives off streams of intensely heated air particles.

Tesla proposes the name of hot St. Elmo's fire for

FIG. 105.—LUMINOUS DISCHARGE WITH INTERPOSED INSULATORS.

this form of discharge, which has the appearance shown in Fig. 106.

Since current impulses produced by alternating discharges of such high frequency cannot be passed through conductors of measurable dimensions, Tesla has experimented as to the effects which such discharges produce on refractory insulating materials placed inside of closed vessels by subjecting such

substances to the thrusts of alternating electrostatic
fields. By connecting vessels or globes with sources
of rapidly alternating potential, under the influence
of the alternating electrostatic thrusts, the molecules
of the residual gas are set into motion with enormous velocity, and the molecular shocks thus given
to the refractory material render it highly luminous.
These effects are best obtained in vacuous spaces.

FIG. 106.—HOT ST. ELMO'S FIRE.

In obtaining the effects of luminosity by the bombardment of the molecules of the residual gas it is
not necessary to connect both terminals of the vessel
to the source ; a single connection suffices. In this
manner a new species of electric lamp is produced,
which may be called the electric bombardment lamp.

Tesla has constructed a great variety of electric
bombardment lamps, in which single or double fila-

ments are employed connected to either one or both
terminals of a rapidly alternating source.

When an electric bombardment lamp is operated
by connecting it to a single terminal only, its brilliancy is greatly increased by connecting one terminal
to a conducting surface placed on the outside of the
lamp and acting as a condenser coating, and connecting the other terminal to a solid body of the same
extent of surface as the condenser coating.

When electric bombardment lamps are provided
with an external coating, they can be lighted when
suspended from a single terminal in any part of the
room. In some cases Tesla succeeded in lighting a
lamp by placing it anywhere in the rapidly alternating electrostatic field obtained between two extended plates of metal. Here no connection of either
terminal to the lamp is necessary, and the lamp remains lighted as it is carried about in different parts
· of the room.

In order to obtain extraordinarily high frequencies,
Tesla adopted the plan of employing the discharges
obtained as above from the secondary of the induction coil to charge a condenser, the discharges of
which were employed to feed the primary of a sec·
ond induction coil, the secondary of which was connected to the circuit of the lamps to be lighted.

The general arrangement of the apparatus employed by Tesla for this purpose is shown in Fig.
107. *G*, is a dynamo producing alternating currents
of comparatively low potential but high frequency.

In this circuit is placed the primary coil *P*, of
an induction coil, which induces alternating currents
of high potential in the secondary circuit *S*.

These currents are employed for charging the condenser *C*, which is provided with an air gap at *A*, and

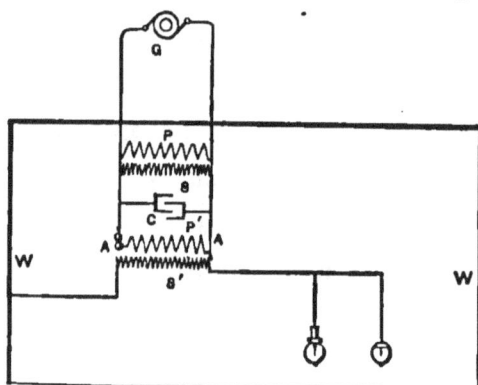

FIG. 107.—TESLA'S HIGH-FREQUENCY CURRENTS SYSTEM OF
LIGHTING.

another primary coil *P'*. The discharges of the condenser across the air gap produce oscillatory or
alternating currents of enormous frequency, which
in passing through the primary *P'*, produce similar
currents, but of very high potential, in the secondary
coil *S'*.

Two incandescent electric lamps, as shown in the

figure, are connected to one pole of the secondary
circuit; one, an incandescent-ball lamp, and the
other a single straight-filament lamp. The other
pole of the secondary circuit is connected to a large
plate *W W*. The construction of these two lamps is
shown respectively in Figs. 108 and 109.

In Tesla's incandescent-ball electric bombardment
lamp, electrostatic waves of high frequency of alter-

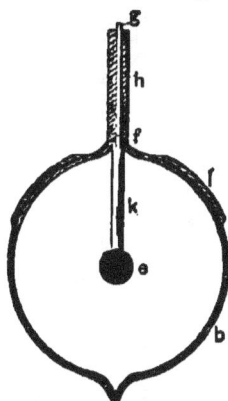

FIG. 108.—TESLA'S INCANDESCENT-BALL ELECTRIC LAMP.

nation, acting on a sphere or ball of carbon connected
with a single filament, as shown in Fig. 108, and
placed inside the vacuous space of a glass chamber,
render such ball or sphere incandescent.

In Tesla's straight filament incandescent lamp, the
carbon ball is replaced by a straight filament of
carbon placed inside an exhausted glass chamber. The
filament is rendered highly luminous on being ex-

posed to electrostatic thrusts or waves of high frequency.

The glass globe *b*, Fig. 109, of the lamp, is provided with a cylindrical neck, inside of which is
placed a tube *m*, of conducting material, on the side
and over the end of the insulated plug *n*.

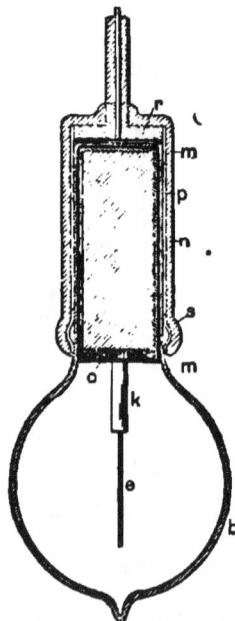

FIG. 109.—TESLA'S STRAIGHT-FILAMENT INCANDESCENT LAMP.

The light-giving filament *e*, is a straight carbon
stem, connected to the plate by a conductor covered
with a refractory, insulating material *k*. An insulated tube-socket *p*, provided with a metallic lining
s, serves to support the lamp and connect it with

one pole of the source of alternating discharges. The other terminal of the machine may be connected to the metal-coated walls of the room, or to metallic plates suspended from the ceiling.

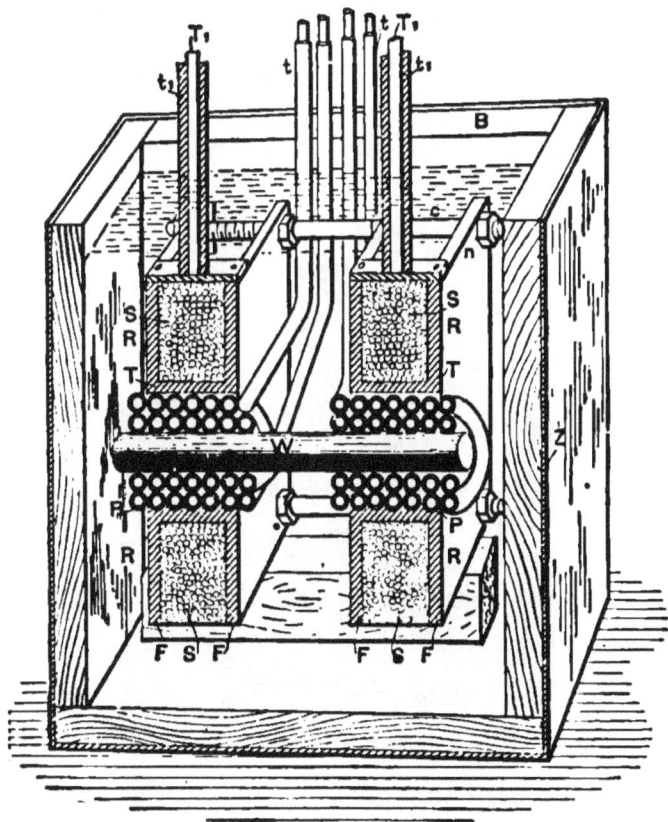

FIG. 110.—DISRUPTIVE DISCHARGE COIL.

When condensers are employed to charge the primary of a transformer. Tesla employed for certain experiments the transformer shown in Fig. 110, in which the box *B,* of hard wood, is covered on the outside with zinc. The coils consist of spools of hard rubber wound with gutta-percha covered wires, *P P,* and *S S,* which form the primaries and secondaries respectively. In order to avoid the effects of the

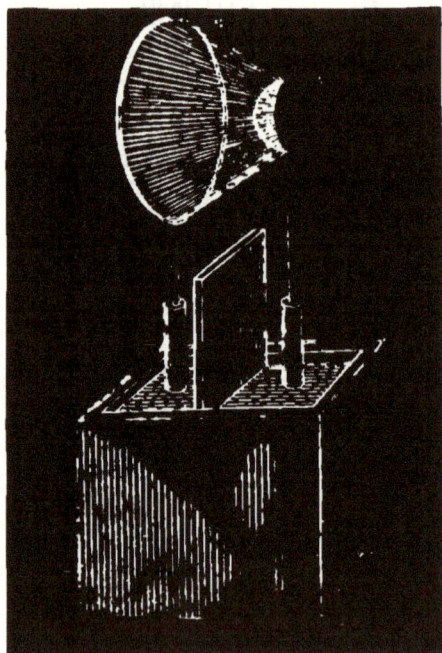

FIG. 111.—DIRECTED STREAMING DISCHARGE.

metal covering of the box the coils are placed as nearly at the centre of the box as possible. Both the primary and secondary coils are wound on the spools in two equal parts. The box is filled with oil from which all air or gas has been removed by boiling.

When the conditions are such in the operation of such a coil that a streaming discharge has been obtained, by connecting the terminals, as shown in Fig. 111, a hollow continuous luminous cone is obtained between the electrodes *W* and *S.*

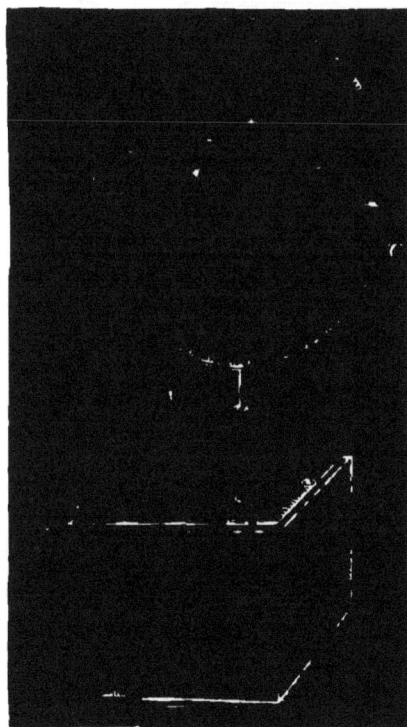

FIG. 112.—LUMINOUS DISC-SHAPED DISCHARGE.

In a similar manner, if the terminals be shaped in the form of rings that are placed consecutively in as nearly the same plane as possible, as shown in Fig.

112, the entire space between the concentric electrode is filled by a luminous discharge.

Probably the most curious form of discharge obtained by Tesla among the many other remarkable

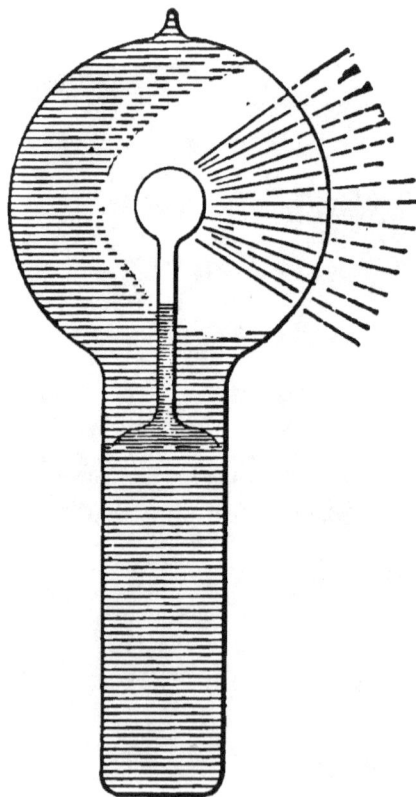

FIG. 113.—ROTATING-BRUSH DISCHARGE.

forms is what he calls the rotating-brush discharge,
and is shown in Fig. 113.

This discharge is taken in a lamp chamber or bulb

containing a high vacuum. A barometer tube *b*,
Fig. 114, is placed inside the chamber and is blown
out into a small bulb, *s*, at one end, and is placed as
shown in Figs. 113 and 114. Under certain con-

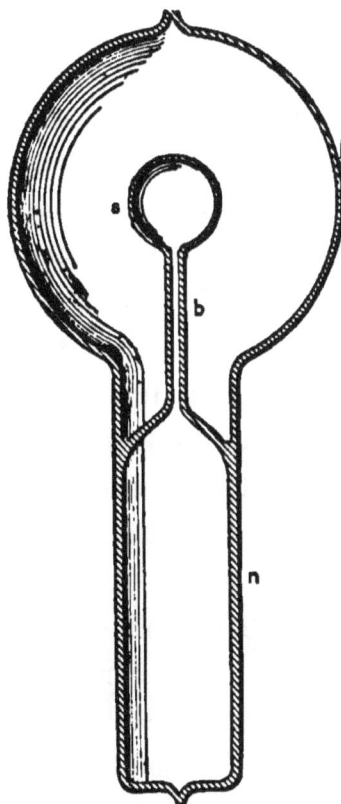

Fig. 114.—Apparatus for Rotating-Brush Discharge.

ditions a brush discharge is obtained that is exceedingly sensitive to electrostatic or magnetic influ-
ences. If the bulb is attached to a single terminal

at its lower end, and is hanging vertically downward, the mere approach of an observer will cause the brush to fly to the opposite side, and, if he walks around the bulb, it will move with him, always keeping on the opposite side. In the Northern Hemisphere the rotation is invariably clock-wise.

Elihu Thomson obtains discharges of high frequency and enormous difference of potential by employing discharges of high potential from a condenser, to produce electro-dynamic induction in induction coils. The high insulation necessary for separating the primary from the secondary coils in such induction apparatus is obtained by surrounding them with oil. By these means he obtains discharges through thirty inches of air at differences of potential that have been estimated to be as high as 500,000 volts.

In his apparatus Thomson employs Leyden jars as condensers. The circuit connections between the jar and the induction coil will be understood from an inspection of Fig. 115 and the following description.

The construction of the apparatus employed by Thomson is described by him as follows :

" A double coil was made, of which the inner turns were about twelve and the outer turns twenty.

These were kept separated from each other, and a branch wire taken from the line and slid from point to point on the outer wire enabled the effective length of the same to be adjusted. The inner coil was connected through a small spark gap, as at *A*, to the outer coating of a Leyden jar, while the wire *L*, was brought near the pole of the jar, which was continuously being charged from a Töpler-Holtz machine

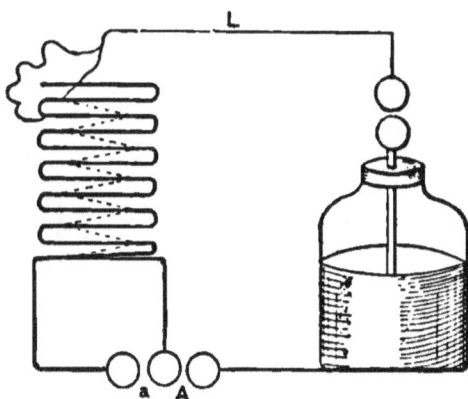

FIG. 115.—THOMSON'S APPARATUS.

the discharge, in passing from the knob of the jar to the wire *L*, representing the line passed by the inner coil. When a certain length of the outer coil was employed only a very short, almost imperceptible, spark was obtained at *A*. If the balance of the turns were disturbed by including more or less than the proper number of the outer turns, not only did a vigorous spark occur, but the gap at *A* could be quite considerably extended, in accordance with the amount of departure taken from the proper number of turns required to produce the balance."

When the apparatus is arranged as shown in Fig. 116, curious three-branched sparks are obtained, which, under certain circumstances, assume remark-

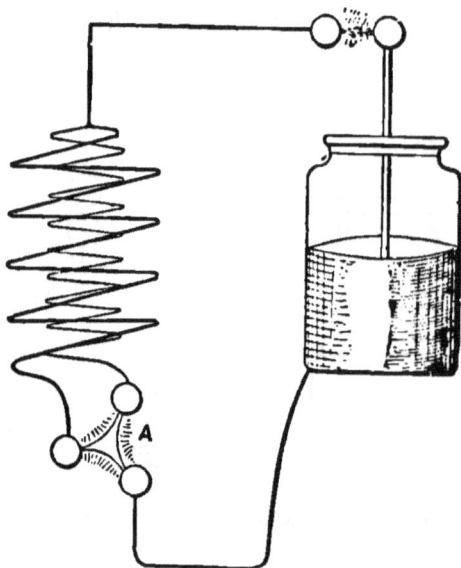

FIG. 116.—ARRANGEMENT OF APPARATUS FOR THREE-BRANCHED SPARKS.

able T and Y shapes. One of these shapes is shown in Fig. 117.

As regards the physiological effects of shocks produced by means of alternating currents, it can readily be shown that up to a certain rate of frequency these shocks are more severe and painful in the case of alternating discharges than in the case of steady currents of equal volume and potential difference. When, however, the rate of alternation increases and reaches a certain limit, the effects of the alternating discharge are much less severe than those of a steady current.

Tesla has shown, beyond any doubt, that when the rapidity of alternation increases beyond a certain extent, the rapidly alternating currents produced are much less dangerous in their effects than a low

FIG. 117.—THREE-BRANCHED SPARKS.

frequency discharge of the same difference of potential. Speaking of this fact before the American Institute of Electrical Engineers on the 20th of May, 1891, Tesla says :

"I have found that by using the ordinary low frequencies the physiological effect of the current required to maintain at a certain degree of brightness a tube four feet long, provided at the ends with outside and inside condenser coatings, is so powerful that I think it might produce serious injury to those not accustomed to such shocks; whereas with 20,000 alternations per second the tube may be maintained at the same degree of brightness without any effect being felt.

" This is due principally to the fact that a much smaller potential is required to produce the same light effect, and also to the higher efficiency in the light production."

Dr. Tatum has made a number of experiments in this direction, and reaches practically similar conclusions.

EXTRACTS FROM STANDARD WORKS.

In "Modern Views of Electricity,"* Lodge, on page 284, says concerning the manufacture of light:

The conclusions at which we arrived, that light is an electrical disturbance, and that light-waves are excited by electric oscillations, must ultimately, and may shortly, have a practical import.

Our present systems of making light artificially are wasteful and ineffective. We want a certain range of oscillation, between seven thousand billion and four thousand billion vibrations per second. No other is useful to us, because no other has any effect on our retina ; but we do not know how to produce vibrations of this rate. We can produce a definite vibration of one or two hundred or thousand per second ; in other words, we can excite a pure tone of definite pitch, and we can command any desired range of such tones continuously by means of bellows and a keyboard. We can also (though the fact is less well known) excite momentarily definite ethereal vibrations of some million per second, as I have explained at length ; but we do not at present seem to know how to maintain this rate quite continuously. To get much faster rates of vibration than this we have to fall back upon

---

* "Modern Views of Electricity," by Oliver J. Lodge, D. Sc., LL. D., F. R. S. London: Macmillan & Co. 1889. 480 pages, 67 illustrations. Price, $2.

atoms. We know how to make atoms vibrate. It is done by what we call "heating" the substance, and if we could deal with individual atoms unhampered by others it is possible that we might get a pure and simple mode of vibration from them. It is possible, but unlikely ; for atoms, even when isolated, have a multitude of modes of vibration special to themselves, of which only a few are of practical use to us, and we do not know how to excite some without also the others. However, we do not at present even deal with individual atoms. We treat them crowded together in a compact mass, so that their modes of vibration are really infinite.

We take a lump of matter, say a carbon filament or a piece of quicklime, and by raising its temperature we impress upon its atoms higher and higher modes of vibration, not transmuting the lower into the higher, but superposing the higher upon the lower, until at length we get such rates of vibration as our retina is constructed for, and we are satisfied. But how wasteful and indirect and empirical is the process. We want a small range of rapid vibrations, and we know no better than to make the whole series leading up to them. It is as though, in order to sound some little shrill octave of pipes in an organ, we were obliged to depress every key and every pedal, and to blow a young hurricane.   .

XII.—ELECTRO-DYNAMIC INDUCTION.

The general principles according to which electricity is produced by the aid of magnets was discovered by Faraday in 1831.

In order to obtain electricity by the aid of a magnet it is only necessary that a conductor be so brought into or moved through the field of the magnet as either to cut or to be cut by its lines of force.

When a conductor cuts, or is cut by, such lines of force, differences of potential are produced in it, and, if such points or places of difference of potential are connected by a conductor, so as to complete a circuit, electric currents are produced.

The production of electromotive force in this manner by cutting lines of magnetic force is called electro-dynamic induction.

We have already seen that when a current of electricity flows through a conductor, a magnetic field, traversed by lines of magnetic force, is produced in the space around the conductor. Such a magnetic field, like the field produced by a magnet, can be employed for the production of electro-dynamic induction.

When an electric current passes through a conductor lines of magnetic force are produced in the space around the conductor. As the strength of the current through the conductor increases, the lines of force of the field increase in number and expand or move outward from the conductor. As the strength of the current decreases, the lines of force decrease in number and contract or move inward toward the conductor.

Even if a conductor be fixed it will have differences of potential induced in it if it is placed in the neighborhood of another conductor, through which a current is passing that is rapidly undergoing changes in its strength ; for, the contracting and expanding lines of force of the latter will cut the neighboring conductor and produce currents in it in one direction as they pass through it on expanding, and in the opposite direction as they pass through it on contracting.

Electro-dynamic induction may, therefore, be produced in two ways :

(1.) By moving a conductor across lines of magnetic force so as to cut these lines ; or,

(2.) By causing expanding or contracting lines of force to pass across a stationary conductor.

Four cases of electro-dynamic induction may arise ; namely,

(1.) Self-induction or inductance; or that form of electro-dynamic induction in which the expanding and contracting lines of magnetic force, produced by varying the strength of the current in any circuit, are caused to pass across or cut that circuit and thus produce differences of potential therein. Self-induction occurs especially in coils.

(2.) Mutual induction or voltaic current induction; or that form of electro-dynamic induction in which the contracting and expanding lines of magnetic force, produced in one circuit by varying the current strength in a neighboring circuit, are caused to pass across another neighboring circuit and thus produce differences of potential therein.

(3.) Magneto-electric induction; or that form of electro-dynamic induction in which a conductor is so moved across the field of a permanent magnet as to cut its lines of magnetic force; or, what is the same thing, in which a permanent magnet is moved past a conductor so as to cause its lines of magnetic force to pass across the conductor and thus, in either case, to produce differences of potential in the conductor.

(4.) Electro-magnetic induction, or that form of electro-dynamic induction in which a conductor is so moved through the field of an electro-magnet, or in which an electro-magnet is moved past a conductor, so as to ensure a cutting of its lines of magnetic force, and thus produce differences of potential therein.

Magneto-electric and electro-magnetic induction are in reality one and the same variety of electro-dynamic induction. They are, therefore, sometimes called dynamo-electric induction.

Self-induction.—When the circuit of a single voltaic cell is closed by connecting its terminals together without including anything else in the circuit, no very intense spark is observed, either when such terminals come into contact, or when such contact is broken. If, however, the terminals are connected to a comparatively long coil of insulated wire, although no appreciable spark is observed on making or closing the circuit, quite a considerable spark is seen on breaking or opening the circuit.

The cause of the increased length of spark thus produced, on opening or breaking the circuit of the cell, is as follows :

When, on the closing of the circuit, the current strength is increasing from zero, or no strength, to the full strength the cell is able to produce, the magnetic field of the coil is increasing, and its lines of magnetic force are expanding or moving outward.

When, on the breaking or opening of the circuit, the current strength in the circuit is decreasing, the lines of magnetic force are contracting or moving inward toward the conductor.

Now, the lines of force moving either from or toward any portion of the circuit or conductor, on expanding or contracting will pass through or cut some other portion of the wire, and will thereby produce differences of potential therein.

As soon as the current strength in the conductor becomes constant its lines of magnetic force no longer move inward or outward, and no effects of electro-dynamic induction are produced. It is, therefore, necessary in these varieties of electro-dynamic induction to rapidly make and break the circuit.

When the circuit is closed, during the time the current strength is increasing, the induced current tends to flow in a direction opposite to that of the inducing current. When the circuit is opened, during the time the current strength is decreasing, the induced current tends to flow in the same direction as the inducing current.

The two currents produced respectively on the making and the breaking of the circuit are called extra currents. That produced in making the circuit is called the extra inverse current, because it flows in the opposite direction to the inducing current ; and that produced on breaking the circuit is called the extra direct current, because it flows in the same direction as the inducing current.

The reason no spark is observed on making the circuit of a long coil, to which the terminals of a single voltaic cell are connected, will now be understood. The differences of potential, produced on the closing of the circuit, are such that the current thereby generated tends to flow in the opposite direction to that of the original current, and thus decreases its strength ; while those produced on the breaking of the circuit cause a current that flows in the same direction, and, therefore, prolongs and strengthens the original current.

Mutual induction is caused, in a similar manner, by the expanding and contracting lines of force, which are produced by rapidly varying the strength of current passing in one circuit, cutting or passing through a neighboring circuit and so producing differences of potential therein. These differences of potential tend to produce currents in one direction when the lines of force are expanding, and in the opposite direction when they are contracting.

The effects of mutual induction can be readily shown by means of the apparatus represented in

Fig. 118, where *B*, consists of two concentric coils of insulated wire that are separately wound on a hollow core of vulcanite or other insulating material. One of these coils is called the primary coil, and the other the secondary coil. The terminals of the primary coil are connected to the poles *P* and *N*, of a voltaic cell. The terminals of the secondary coil are connected to the galvanometer *G*.

If, now, the circuit of the voltaic cell be closed through the primary coil, then at the moment of

FIG. 118.—MUTUAL INDUCTION.

closing the circuit, a current is produced in the secondary coil, which flows in the opposite direction to the current flowing through the primary, as is indicated by the deflection of the needle of the galvanometer in a certain direction. If the circuit be opened, then at the moment of opening, a current is produced in the secondary, which flows in the same direction as the current in the primary, as is indicated by the deflection of the galvanometer needle in the opposite direction.

As in the case of self induction these currents are but momentary, and continue only as long as the current in the primary varies in strength.

When the current strength is fully established in the primary coil, and no current exists in the second-ary, if a short circuit is formed across the battery terminals by placing the wire d, in the mercury cups x and y, shown in Fig. 119, the current in the pri-mary is thereby decreased and a current is induced in the secondary coil in the same direction as that flowing in the primary circuit.

FIG. 119.—MUTUAL INDUCTION.

In *magneto-electric induction*, the current is ob-tained by means of differences of potential estab-lished in a conductor, either by moving the conduc-tor through the field of a permanent magnet, so as to cut its lines of force, or by moving the magnet past the conductor so as to cause the lines of force to pass across the conductor.

Magneto-electric induction can be readily shown

by means of the apparatus represented in Fig. 120. The terminals of a coil of insulated wire are connected to the terminals of a galvanometer. When the magnet *M*, is moved toward or from such coil the needle of the galvanometer will be deflected by a current, which will flow in one direction as the magnet approaches the coil, and in the opposite direction as it moves away from it.

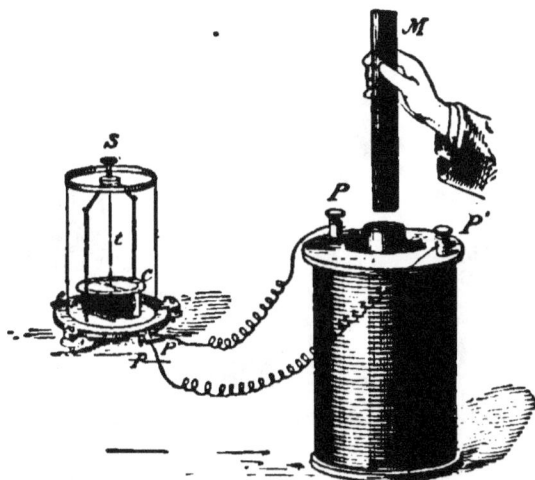

FIG. 120.—MAGNETO-ELECTRIC INDUCTION.

In *electro-magnetic induction*, a coil or conductor, moved through the field of an electro-magnet so as to cut its lines of force, has a current produced in it by the differences of potential thereby generated in the coil or conductor. This kind of induction differs from magneto-electric induction only in the fact that the magnet used is an electro-magnet and not a permanent magnet.

Electro-magnetic induction can be shown by means of the apparatus represented in Fig. 121. The terminals of a voltaic cell are connected to the ends of a coil of insulated wire, as shown. The terminals of a second hollow coil, provided with a sufficiently wide opening to permit the first coil being moved

FIG. 121.—ELECTRO-MAGNETIC INDUCTION.

into and out of it, are connected to the terminals of a galvanometer. When the coil connected to the battery—which may be called the primary coil—is moved into or out of the coil of wire whose terminals are connected to the galvanometer—which may be called the secondary coil—currents are produced in it, as is shown by the deflection of the needle of the galvanometer. These currents flow in one direction as the primary circuit is moved toward the secondary circuit, and in the opposite direction as it is moved away from it.

The following considerations will show that the production of currents by electro-magnetic induction is in reality the same as their production by magneto-electric induction.

When a steady current is flowing through a coil of wire placed in the neighborhood of another coil of wire, no difference of potential is produced in the neighboring wire as long as such coils remain fixed. If, however, either be moved toward the other, so that the lines of force produced by one coil are caused to pass through the other coil, differences of potential will be produced, and the direction of the resulting current will be opposite to the current flowing through the inducing coil when they are moved toward one another, and in the same direction when they are moved from one another.

In the same way, when a permanent magnet is moved toward a conductor, or a conductor is moved toward a permanent magnet, so as to cause the lines of force to move across the conductor, differences of potential are produced, which are respectively inverse and direct as the one approaches or moves away from the other, and their directions may be deduced by reference to the direction of the Ampèrean currents, which are assumed to produce the magnetism of a permanent magnet.

Magneto-electric induction and electro-magnetic induction are, therefore, sometimes called dynamo-electric induction.

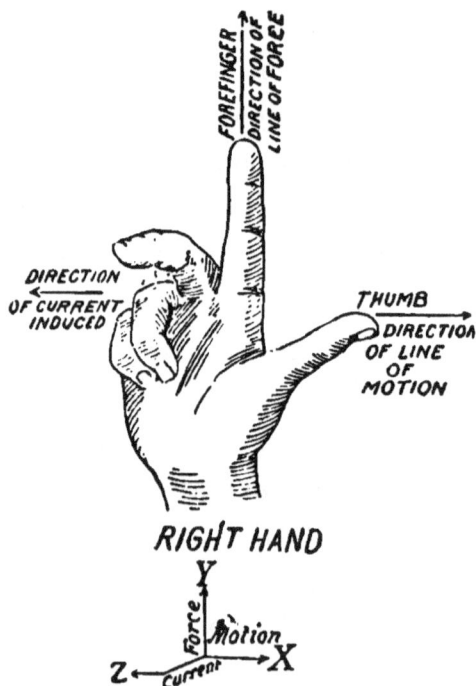

FIG. 122 —FLEMING'S RULE.

The same principles may be expressed by the following laws :

(1.) Any increase in the number of lines of magnetic force which pass through a loop of a circuit, produces an inverse current in that circuit; any decrease in the number of such lines produces a direct current in that circuit.

(2.) The intensity of the induced current, or, more correctly, the difference of potential produced, is proportional to the rate of increase or decrease of the lines of force passing through the loop.

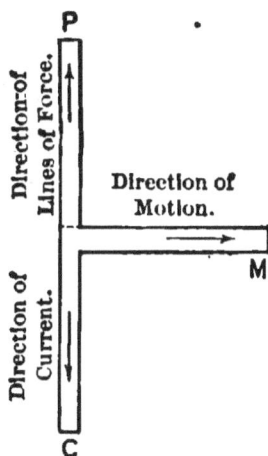

FIG. 123.—FLEMING'S RULE.

The direction of the currents produced by dynamo-electric induction may be remembered by the following plan suggested by Fleming. Let the right hand be held with the fingers extended as shown in Fig. 122. Let the forefinger represent the positive direction of the lines of force—that is, as coming out of the north pole of the magnet ; then if a conductor be moved in the direction in which the thumb points it will have a current produced in it by induction, which will flow in the direction in which the middle finger points.

Or, the same thing can be more readily remembered by cutting a piece of paper in the shape shown in Fig. 123. Marking it as shown, and bending the arm upward at the dotted line, so as to form three axes at right angles to one another, then, if the arm *P*, represents the direction of the lines of force, a conductor, moved in the direction of the arm *M*, so as to cut these lines at right angles, has a current produced in it which will flow in the direction of the arm *C*.

EXTRACTS FROM STANDARD WORKS.

In a work entitled " The Alternate Current Transformer in Theory and Practice,"* by J. A. Fleming, the following description is given on page 36, Vol. I., concerning some of the phenomena of electrodynamic induction:

As soon as we cease to limit our consideration to constant or steady currents we find that we shall not be able to give a full account of the phenomena unless we extend our ideas and recognize another quality of conductors equally important with resistance in determining the numerical ratio of instantaneous current strength to instantaneous potential difference between two points in any linear conductor traversed by that current. This quality of the circuit is called its *Inductance.*

The clear recognition of this special quality of a conductor dates from the publication of Faraday's memoir forming the Ninth Series of his " Electrical Researches " (§1,048 1st Ed.), *On the Influence by Induction of an Electric Current on Itself*, and from the investigations of Prof. Joseph Henry (*Phil. Mag.*, 1840), of Princeton. The chain of experiments which lead to these ideas was apparently started by the inquiry addressed to Faraday by a Mr. Jenkin one

---

* "The Alternate Current Transformer in Theory and Practice," by J. A. Fleming, M.A., D. Sc. 2 vols. London: Electrician Printing and Publishing Company. 1889-92. Vol. I. The Induction of Electric Currents. 487 pages, 157 illustrations. Price, $3. Vol. II. The Utilization of Induced Currents. 591 pages, 300 illustrations. Price, $5.

Friday evening, at the Royal Institution, as to the reason why a shock was experienced when a circuit containing an electro-magnet was broken, the observer retaining in his two hands the ends of the circuit, but no shock was felt if the circuit contained neither magnet nor long coils of wire. Faraday seems speedily to have arranged an organized attack on the subject and to have returned from his investigation burdened with the spoils of victory in the shape of the following facts :

(1.) If a battery circuit is closed by a short thick wire, then, although there may be a strong current existing in this wire on breaking contact at any point, little or no spark is seen, and if the two ends of the circuit are grasped in the two hands, and the interruption takes place between the hands, then little or no shock is experienced.

(2.) If a long wire is used instead, then, although the absolute strength of the current may be less, yet the spark and shock at interruption are more manifest.

(3.) If this length of insulated wire is coiled up in a helix on a pasteboard tube, then, although the length of the wire and the strength of the current are the same, yet the spark and the shock are still more marked.

(4.) If the above helix has an open iron core placed in it, both these effects are yet more exalted.

(5) If the same length of wire is doubled on itself, being, however, insulated, then the effects nearly vanish, and, whether straight or coiled, this double wire with current going up one side and down the other is no better in respect of spark and shock on interruption than a very short wire.

XIII.—INDUCTION COILS AND TRANS-
FORMERS.

An induction coil consists of any arrangement of parts by means of which electric currents may be produced by that variety of electro-dynamic induction called mutual induction.

In an induction coil alternating currents passed through a conductor arranged in the form of a coil, induce alternating currents in a neighboring coil of wire.

The coil of wire through which the alternating currents are passed is called the primary coil. The coil in which the alternating currents are produced is called the secondary coil.

As the current strength in the primary coil or conductor varies, it produces a field whose lines of magnetic force expand or move outward from the conductor while the current strength increases, and contract or move inward toward the conductor while the current strength decreases. In these movements the lines of force cut the conductor of the secondary coil, either during their motion toward or from it, and produce therein differences of potential, which cause electric currents to flow through the secondary circuit.

The directions of the currents produced in the secondary circuit are as follows :

(1.) Opposite to the currents in the primary coil, on the making of the circuit, or as the current is increasing, and the lines of magnetic force cut the secondary circuit on expanding.

(2.) In the same direction as the currents in the primary coil, on the breaking of the circuit, or as the current is decreasing, and the lines of magnetic force cut the secondary circuit on contracting.

The rate of alternation of the current in the secondary coil is the same as that in the primary.

The relative values of the difference of potential produced in the secondary as compared with that in the primary depend on the relative lengths of the wires in the primary and secondary coils. If, for example, the length of wire in the secondary coil circuit is fifty times the length of that in the primary, the difference of potential in the secondary will be fifty times that of the primary. If, on the contrary, the length of wire in the secondary coil is shorter than that in the primary, the difference of potential in the secondary will be less than that in the primary. If, for example, the secondary has a length one-fiftieth of that of the primary, the difference of potential in the secondary will be one-fiftieth of that of the primary.

Since in a well-constructed induction coil there is very little energy lost in producing these differences of potential by mutual induction, it will be readily understood that the amount of energy produced in the secondary will be very nearly equal to the amount expended in the primary.

Assuming that the energy expended in the primary in the form of electric work may be approximately expressed as $C\,E$, in which C, equals the the current in ampères, and E, the electromotive force of the primary in volts; then, when there is no loss of energy in an induction coil. $C\,E = C'\,E'$.

From the above expression it will be seen that in whatever proportion the difference of potential is increased in the secondary, by using more turns of wire, in that proportion must the strength of the secondary current be decreased ; since otherwise the product of C' and E', would not be equal to the product of C and E. If, therefore, the difference of potential is increased in the secondary the current strength in the secondary must be proportionally decreased.

If, on the contrary, the difference of potential in the secondary be decreased by using fewer turns of wire, so far then must the current strength of the secondary be increased.

There thus arise two distinct varieties of induction coils, namely :

(1.) Those in which the wire in the secondary coil is longer than that in the primary, and in which the difference of potential induced in the secondary is, therefore, greater than that in the primary, and the current strength in the secondary is less than that in the primary.

(2.) Those in which the length of wire in the secondary coil is shorter than that in the primary coil, and in which the difference of potential induced in the secondary is therefore smaller than that in the primary, and the current strength in the secondary is greater than that in the primary.

Originally, induction coils were made of the first form only ; when, therefore, they came to be made of the second form—namely, in which the length of the conductor of the secondary is shorter than that of the primary—they were called inverted induction coils.

A well-known form of induction coil, named after its inventor the Ruhmkorff coil, belongs to the first class, or that in which the secondary wire is of much greater length than the primary.

The general construction of this form of coil is shown in Fig. 124. The primary conductor consists of but a few turns of thick wire, wound on a core formed of a bundle of soft iron wires, and having its ends brought out at f, f.

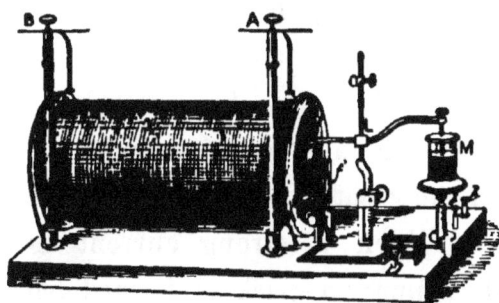

FIG. 124—RUHMKORFF COIL.

The resistance of the primary coil is made low in order that a strong current may be passed through it.

The long, thin wire forming the secondary is wrapped on the surface of a cylinder of hard rubber or glass surrounding the primary coil. In some coils this wire is more than one hundred miles in length.

The primary coil will induce more powerful effects in the secondary if it is provided with a core of iron. In order to prevent much of the energy of the primary from being wasted in producing induction currents in this core it is made of soft iron wires.

When the primary circuit is rapidly made and broken, alternating currents are produced in the secondary of great electromotive force but of small current strength.

The making and breaking of the primary circuit, in the form of Ruhmkorff coil shown in the figure, is effected by means of a conductor dipping into a mercury cup, as shown at *M*.

The more suddenly the primary circuit is made and broken the greater is the effect produced in the secondary. When a strong current is passing through the primary a spark is formed between the contact points, at which the primary circuit is made and broken. This spark prolongs the duration of the primary current and thus decreases its efficiency. In order to prevent the formation of this spark a condenser is generally connected to the primary circuit in the manner shown in Fig. 125, which represents diagramatically the arrangement of the different parts of an induction coil. The core consists of a bundle of soft iron wires as shown at . *I, I'*. The primary wire *P P*, consists of a few turns of thick wire, while the secondary *S S*, consists of many turns of fine wire. In order, however, to

prevent confusion of details, the secondary is represented in the figure as consisting of but a few turns
of wire.

The terminals of the battery *B*, are connected to
the primary circuit in the manner shown. The
automatic contact-breaker—provided for obtaining
the rapid alternations of the primary circuit—is
shown at *H* and *O*. The piece of soft iron *H*, is

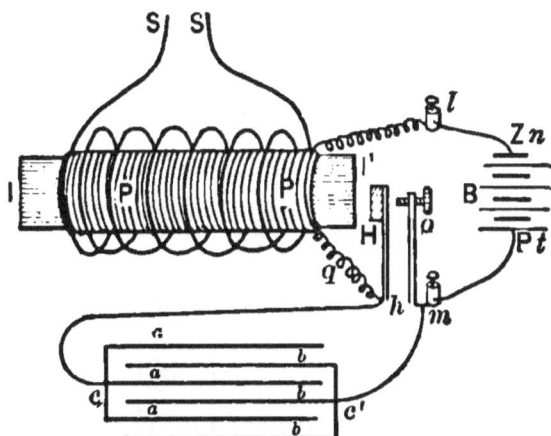

FIG. 125.—CIRCUIT CONNECTIONS OF INDUCTION COIL.

supported on the metal spring *h*, near the soft iron
core *I'*. When no current is passing through the
primary circuit—that is, when it is open or broken—
the spring rests against the platinum contact point
O; but when the circuit is closed, the battery current flows through the primary coil, and magnetizes
the core *I, I'*, which then attracts the iron piece *H*,

thus breaking the circuit of the battery. But this in turn causes the core *I*, to regain its magnetism, and therefore the piece *II*, again comes to rest with the spring resting against the contact point *O*, and in this way a rapid to-and-fro motion of the mass *II*, is obtained, or a rapid making-and-breaking of the battery circuit.

The principles involved in the production of currents by mutual induction were discovered by Prof. Henry, who was the first to point out the true cause of the extra spark produced on breaking the circuit

FIG. 126.—HENRY'S INDUCTION COILS.

of a comparatively long coil of insulated wire. Henry also showed that the current from the secondary of one induction coil could be passed through the primary of another, and so on, thus intensifying the effects.

A series of three of Henry's induction coils is shown in Fig. 126. An alternating current sent through *a*, will produce induced currents in its secondary *b ;* now, *b*, is connected in series with another primary coil *c*, whose secondary *d*, is similarly

connected to another primary *e.* The currents induced in its secondary *f*, are finally employed to
magnetize a piece of iron wire *g*, or are used for any
other desired effect.

Henry called the currents produced at *b*, secondary
currents, or currents of the second order; those produced at *d*, tertiary currents, or currents of the
third order; those produced at *f*, he called currents
of the fourth order, and so on.

In Fig. 127 are represented two such coils arranged for giving a person an electric shock on
grasping the handles at *e* and *f*.

FIG. 127.—HENRY'S INDUCTION COILS.

Since an induction coil converts or transforms one
form of electric energy into another it is often called
a converter or transformer. The latter name is the
one most frequently employed. Although the term
transformer is more generally used in connection
with the inverted induction coil, yet it is now often
applied to induction coils like the Ruhmkorff, in
which the length of the secondary greatly exceeds
that of the primary.

In order to distinguish the two forms of induction coils or transformers the one which increases the electromotive force is called a *step-up transformer*, and the one which decreases the electromotive force is called a *step-down transformer*.

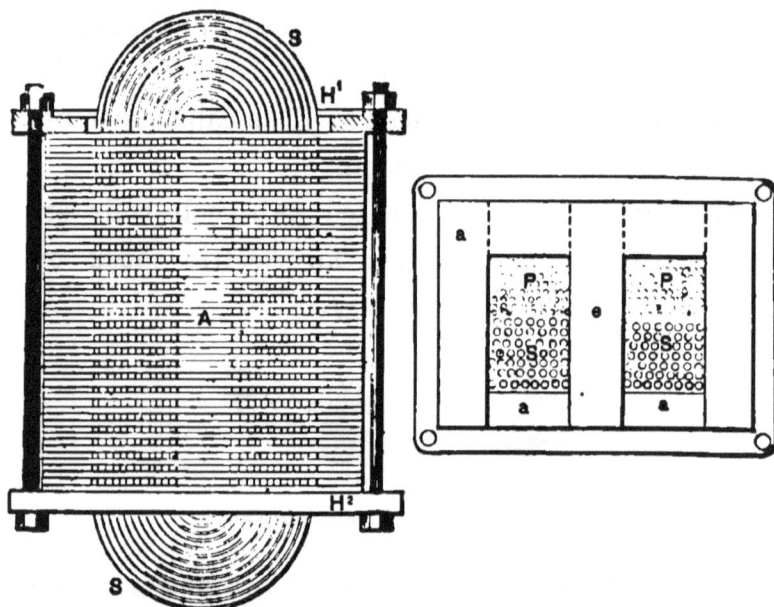

FIG. 128.— CLOSED CIRCUITED TRANSFORMER.

The transformer shown in Fig. 128 is a step-down transformer. It consists essentially of an inverted induction coil in which the primary *P, P,* is formed of many turns of a wire that is thin when compared to the secondary *S, S,* which is formed of a few turns of comparatively thick wire. In order to prevent loss of energy by the production of currents in the core, this part of the transformer is thoroughly laminated. In order to lower the resistance of the magnetic circuit, the transformer shown in the figure is iron-clad.

Step-down transformers are employed in systems of electric light distribution where currents of comparatively small current strength and considerable difference of potential are sent over a line from a distant station into transformers, placed where the electric energy is to be used, by which they are changed into currents of comparatively small difference of potential and considerable current strength.

Transformers may be divided, according to the form of their core, into closed circuited and open-circuited transformers. In the former the iron completely surrounds the coils, in the latter it only partially surrounds them. The iron-clad transformer above described belongs to the closed-circuited class. That shown in Fig. 129, belongs to the open-circuited class. The energy produced in the secondary is somewhat smaller than that expended in the primary on account of the following losses:

(1.) Specific inductive capacity.

(2.) Hysteresis, or magnetic friction.

(3.) Heating of the primary circuit.

(4.) Heating of the secondary circuit.

(5.) Foucault currents.

When a converter is properly constructed the loss of conversion at full load is but small, the number of watts in the secondary being very nearly equal to those in the primary. A current of ten ampères at 2,000 volts, when passed into a converter, the number of whose primary turns is twenty times the number of its secondary turns, will produce a current

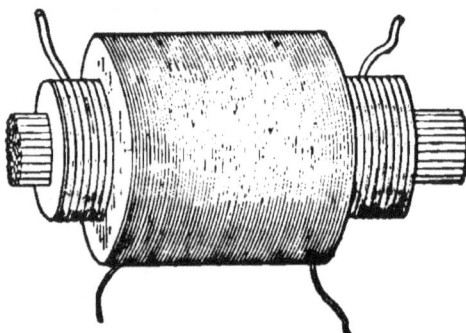

FIG. 129.—OPEN-CIRCUITED TRANSFORMER.

in its secondary whose strength is about twenty times as great, or nearly 200 ampères, but whose voltage is about one-twentieth, or 100 ; the watts in the two cases are nearly the same, or, theoretically, 20,000 watts. In reality it is somewhat smaller.

In a form of apparatus known as the pyromagnetic generator, electric currents are obtained by a species of electro-dynamic induction. Here, how-

ever, the energy of heat is employed to produce alternations in the strength of the magnetism. As is

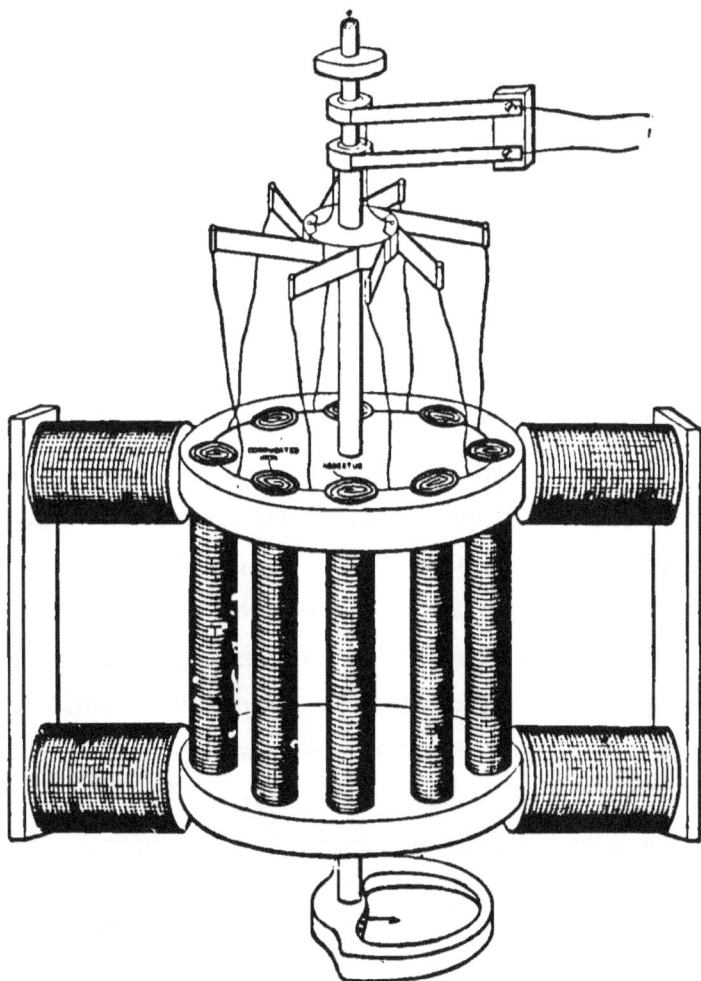

FIG. 130.—PYROMAGNETIC GENERATOR.

well known, the strength of magnetism in iron decreases with an increase of temperature. If, then,

means are provided by which a magnetic mass of
iron is alternately heated and cooled the varying
strength of its magnetism will produce expanding
and contracting lines of magnetic force, which may
be caused to cut a neighboring coiled conductor and
thus produce differences of potential therein.

A pyromagnetic generator is shown in Fig. 130.
This apparatus is sometimes called a pyromagnetic
dynamo. Eight electro-magnets, placed as shown,
are provided each with an armature consisting of a
roll of corrugated iron that has a coil of insulated
wire wound on it, protected by asbestos paper. These
armatures pass through two iron discs, as shown, and
have their coils connected in series in a closed circuit, the wires from the coils being connected with
metallic brushes that rest on a commutator supported
on a vertical axis. A pair of metallic rings is provided to carry off the current generated.

The vertical axis is provided below with a semi-
circular screen called a guard-plate, that rotates
with the axis and cuts off or screens one-half of the
armature from heated air, generated below in a fire.
The difference in the magnetization of the armatures, when hot and cold, produces expanding and
contracting lines of force, which produce electric
currents.

EXTRACTS FROM STANDARD WORKS.·

J. A. Fleming in his "Short Lectures to Electrical Artisans," * on page 75, says as regards induction coils :

Very great precautions as to insulation are essential, in order to obtain long sparks from induction coils. In a large coil, built for the late Mr. Spottiswoode, the secondary coil had a total length of about 280 miles, and the primary a total length of 1,164 yards.

Mr. Spottiswoode obtained very powerful discharges from his coil by disconnecting the interrupter and condenser and sending direct into the primary the alternate currents of a De Meritens alternate current magneto-machine.

Induction coils, such as above, formerly found their applications only in scientific research and experiments, but they have recently, under modifications, become important practical appliances in electric lighting, and in this application are called secondary generators.

An induction coil is a reversible machine. If a current of considerable magnitude circulates under small electromotive force in the primary, then variations in the strength of this give rise to very small currents of exceeding high electromotive force in the secondary. We may reverse this

* "Short Lectures to Electrical Artisans : Being a Course of Experimental Lectures Delivered to a Practical Audience," by J. A. Fleming, M.A., D.Sc. London : E. & F. N. Spon. 1892. 210 pages, 74 illustrations. Price, $1.50.

induction, and cause to circulate in the secondary very small currents under very high electromotive force. These by their fluctuations, will generate in the primary large currents of small electromotive force. We do not, in either case, create electric energy. The energy of a current flowing in a conductor at any instant is measured by the product of the current strength and the electromotive force between the ends of that conductor, and hence electric energy is a quantity which is the product of two factors, current and electromotive force. What the induction coil enables us to do is to increase one of these factors at the expense of the other, and transform our electric energy in form, but not in amount. In this respect we operate on electric energy by means of an induction coil, just as a simple mechanical power, such as a pulley, enables us to operate on mechanical energy, converting a quantity of work which consists of a small stress exerted through a great distance into a large stress exerted through a small distance.

Fleming, in the second volume of the "Alternate Current Transformer,"* speaking of the historical development of the induction coil and transformer, says on page 1:

In following out the stages of development of the induction coil and the transformer, we find that they are no exceptions to the general law that improvements in experi-

* "The Alternate Current Transformer in Theory and Practice," by J. A. Fleming. Vol. II., "The Utilization of Induced Currents." London : The Electrician Printing and Publishing Company. 1892. 591 pages, 300 illustrations. Price, $5.

mental appliances advance along definite lines by a process of evolution in which rudimentary forms are successively replaced by more and more completely developed machines. We are able, by a careful scrutiny of existing and pre-existing modifications, to detect the ideas which at every step have impelled inventions forward, and also to examine the prototypes in their relation to the final and fully developed idea. Most readers would probably consider that the prototypical form of all modern induction coils and transformers was the iron ring with which Faraday made the initial discovery in electro-magnetic induction, and in one sense this is of course correct; but a careful examination of the early stages of the induction coil as we now possess it seems to show that it is descended in a direct line not from Faraday's ring so much as from Henry's flat spirals, and that it is these latter which are the chiefs of the clan and the true ancestors of our modern coil.

Henry's claim to be an independent discoverer of the fundamental fact of electro-magnetic induction is not now disputed. In the July number of *Silliman's American Journal of Science* for 1832, Joseph Henry, then a young teacher in the Albany Academy, gave an account of the manner in which he had independently, and before receiving an account of Faraday's work, performed in the previous autumn, elicited from his own great electro-magnet an induced current by wrapping round the soft iron armature certain coils of insulated wire. In the same paper in which he thus discloses his anticipated discovery he rendered an account of that in which he had in turn anticipated his illustrious rival by the discovery of the fact of the *self-induction* of a spiral conductor, and denoted the phenomena by which it has since been known. Simply confining himself to the bare statement of the new fact that if the poles of a small battery are joined by a wire a foot long no spark will be found on breaking the circuit, whereas if the wire be thirty or forty feet long, and particularly if it be coiled into a spiral, it gives a bright spark when so employed, Henry noted the discovery and correctly attributed the phenomena to the induction of the circuit upon itself. Finding, however, that Faraday was following on the same line of discovery he published in the *Journal of the Franklin Institute,* in March, 1835 (Vol. XV., pp. 169, 170), a brief epitome of the facts he had collected, and made mention, for the first time, of the use of the spirals of flat copper tape or ribbon, insulated and closely wound together, with which he subsequently conducted his brilliant train of discoveries on the mutual induction of circuits.

XIV.—DYNAMO-ELECTRIC MACHINES.

A dynamo-electric machine is any combination of parts by which mechanical energy is converted into electrical energy by means of electro-dynamic induction, that is, by causing conductors to pass through or to cut lines of magnetic force.

The term "dynamo-electric machine" is sometimes applied not only to machines in which mechanical energy is converted into electrical energy, but also to those in which electrical energy is converted by motors into mechanical energy. The term "electric motor," however, is preferable for the latter case, and is now generally so employed. Sylvanus P. Thompson in his "Dynamo-Electric Machinery" defines a dynamo-electric machine as follows :

"A machine for converting energy in the form of mechanical power into energy in the form of electric currents, or vice versa, by the operation of setting conductors (usually in the form of coils of copper wire) to rotate in a magnetic field, or by varying a magnetic field in the presence of conductors."

Originally the term "dynamo-electric machine" was

limited to the case of a machine for converting
mechanical power into electrical power, in which the
machine was self exciting, that is, required no other
current to start it than that produced when its armature was rotated in the permanent field of the machine.

Dynamo-electric machines are now constructed in
a great variety of forms, consisting, however, of the
following parts :

(1.) A rotating part called the armature, which
consists of coils of wire, or conducting bars, strips,
or plates, generally placed on a core of soft iron
and rotated in the magnetic field of the machine so
as to cut its lines of magnetic force. Sometimes the
armature is stationary and the field rotates or pul-
sates.

(2.) The field magnets which produce the magnetic field in which the armature rotates.

(3.) The pole pieces of the field magnets, which
are designed to distribute the field of the field magnets evenly over the rotating armature and to reduce
the resistance of the air gap.

(4.) The commutator by means of which the
currents, produced in the armature by the differences
of potential generated in its conductors on rotation,
are caused to flow in one and the same direction.

In alternating current dynamo-electric machines, in which the currents produced by the armature are not caused to flow in one and the same direction, but flow in alternately opposite directions, this part of the dynamo-electric machine is called the collector, since it merely serves to collect the currents.

FIG. 131.—SERIES DYNAMO.

(5.) The collecting brushes that rest on the commutator cylinder and carry off the current produced in the coils by the differences of potential generated therein on their rotation through the field.

The relations which these different parts bear to one another can be seen in Fig. 131, in which is shown a form of dynamo-electric machine. The field magnets consist of four coils of wire so placed on a heavy core of soft iron, called the field magnet frame, as to produce consequent poles, that is, two north poles, for example, in a massive piece of iron called a pole piece, attached to the frame of the machine and placed at the top of the machine, and two south poles in a similar pole piece placed at the bottom of the machine. These pole pieces are

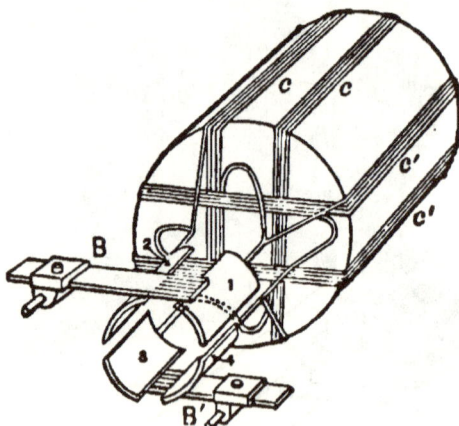

Fig. 132.—Drum-Armature.

placed and shaped so as to conform to the cylindrical outline of the armature which rotates between them.

The armature consists, as shown, of many coils of wire wound around a ring or hollow cylinder, the ends of contiguous coils being connected together and to insulated segments of the commutator on which the collecting brushes rest that carry off the current.

A great variety of shapes may be given to the armatures of dynamos.

A very common form of armature, called the drum-armature, shown in Fig. 132, takes its name from the drum shape of the core on which the wire is wound. As will be seen from an inspection of the

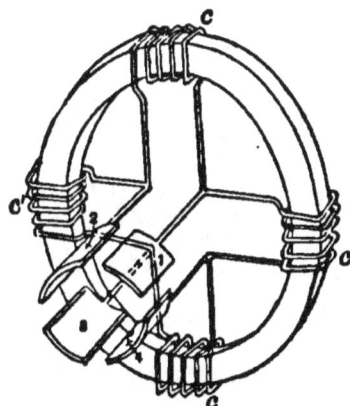

FIG. 133.—RING-ARMATURE.

figure, the armature coils are wound longitudinally over the surface of a closed drum or cylinder, and the ends are afterward connected, in the manner shown, to insulated plates of metal, suitably supported and arranged in the form of a cylinder called the commutator cylinder. In this case the collecting brushes *B*, *B,'* rest on the commutator cylinder in the positions shown, and carry off the current from the armature.

Another form of armature called the ring-armature, from the ring-shape of its core, is shown in Fig. 133. In a ring-armature the coils are connected to one another and to the separate pieces of metal in the commutator cylinder, as shown in the figure; namely, by connecting the beginning and end of contiguous coils together, and to a separate segment of the commutator cylinder. It will be no-

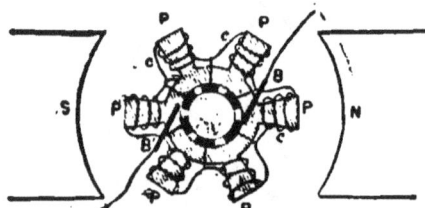

Fig. 134.—POLE-ARMATURE.

ticed that the number of separate parts or segments in a commutator cylinder will depend either on the number of separate coils that are wound on the armature core, or on the number of separate pairs of coils that are first connected together, and afterward connected at their common junction to the commutator segment.

The form of armature shown in Fig. 134 is called a pole-armature. It consists of a series of coils of insulated wire wound on cylindrical cores

that project radially from the periphery of a disc,
drum or ring.

As the armature is rapidly rotated in the magnetic field produced by the two field magnets N and S,
it cuts their lines of magnetic force, and has differences of potential generated in its wires or conductors, that, when such are connected to closed circuits, produce currents which flow in one direction
during motion past one of the magnet poles, and in

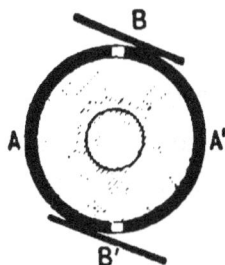

FIG. 135.—COMMUTATOR OF DYNAMO-ELECTRIC MACHINE.

the opposite direction during motion past the other
magnet pole. In order to cause such currents to
flow in one and the same direction they are com-
mutated or changed in their direction by the action
of the commutator.

The operation of the commutator will be understood from an inspection of Fig. 135, which shows
the action that occurs in the case of a coil of wire
rotated between two magnet poles. One end of such

a coil is connected to the insulated segment A', and the other end to the insulated segment A.

The brushes B and B', are so placed on the commutator cylinder that they are in contact with the segments A' and A, respectively, as long as the current flows in the same direction in the coil of wire, but are in contact with A and A', when the current changes its direction, and continue in such contact as long as the current flows in this direction. By these means, therefore, the current will flow in one and the same direction through the circuit connected with the collecting brushes.

Since the commutator segments are subject to wear, both from friction of the brushes and the burning action of destructive sparks, they are generally made of comparatively thick metal; they are insulated from one another and suitably supported, generally by a rocker arm, on the shaft of the armature. The number of metal segments placed on the commutator cylinder will depend on the number, arrangement and connection of the armature coils; it is generally equal to the number of complete coils.

In Fig. 136, a single coil A B, is shown with its ends connected to the two-part commutator C $C,'$ and placed so as to be capable of rotation around the axis R R.

The same connections will be made whether the coil is formed of a single turn or of two or more turns of wire. For example, in Fig. 137 is shown

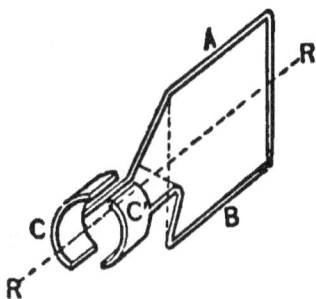

FIG. 136.—CONNECTION OF COIL TO COMMUTATOR SEGMENTS.

a coil *A B*, consisting of two turns similarly connected to the segments of a two-part commutator. So also in Fig. 138 a similar connection is shown of a coil placed on the ring-armature *G*.

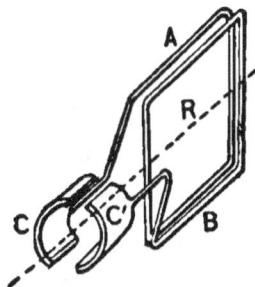

FIG. 137.—TWO-PART COMMUTATOR.

The field magnets of a dynamo-electric machine consist of a frame or core on which the magnetizing coils are wound.

The field magnet cores should be made of thick solid iron as soft as possible, the great size being necessary in order to ensure a powerful magnetic field so as to ensure a high voltage as well as to prevent the magnetizing effect of the armature from too greatly influencing the field of the field magnets.

The pole pieces should also be massive and made of very soft iron. They may, if so desired, be laminated so as to avoid a loss of energy from the produc-

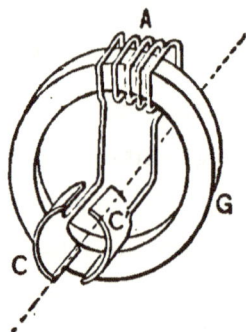

FIG. 138.—CONNECTION OF COIL TO COMMUTATOR SEGMENTS.

tion therein of currents called eddy currents. The pole pieces should preferably extend partly around the armature so as to cause the lines of force of the field magnets to be distributed as much as possible over the armature surface, and to reduce the resistance of the air gap. Care must be taken, however, in bringing the edges of the opposite pole pieces near together, since otherwise the lines of force might pass directly through the air between the edges of the pole pieces, rather than through the armature itself.

The collecting brushes consist of strips of metals, bundles of wire, slotted plates of metal, or plates of carbon that bear on the commutator cylinder and carry off the current.

Various forms are given to the brushes. The commonest, however, are shown in Fig. 139, where

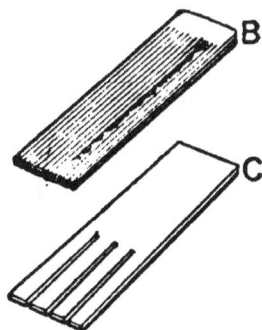

Fio. 139.—BRUSHES.

the brush *B*, is formed of copper wire soldered together at its non-bearing end *B;* that at *C*, is formed of a plate of copper split, as shown, at its non-bearing end. Brushes of carbon are very commonly employed.

Let us now consider a single loop or coil of wire, such, for example, as that shown at *A B C D*, in Fig. 140, supported so as to rotate between the poles *S N*, of the field magnets.

The ends of the coil are connected in the manner shown to the two-part commutator. On the rotation of the coil so that the top of the coil shall move in the direction of the large arrow, differences of potential are generated which will cause an electric current to flow in the direction shown by the smaller arrows during the motion of the loop past the north pole from the bottom to the top. In other words, currents are produced which flow in one direction during one-half of its rotation,

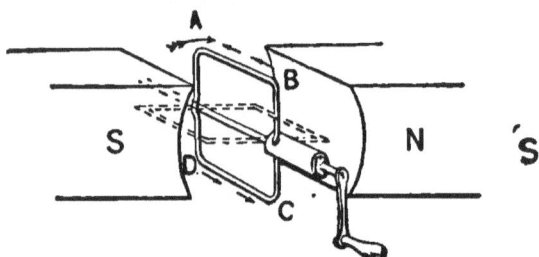

FIG. 140.—INDUCTION IN ARMATURE LOOP.

and in the opposite direction during the other half of its rotation.

If, now, collecting brushes rest on the commutator cylinder in the position shown in Fig. 141, the current will flow in one and the same direction, and B', will become the positive brush because it will be connected with the end of the coil only while it is sending its current into such brush, and since, as soon as the direction of the current in the coil is changed, the other end will by the rotation be moved into connection with the said brush, so that the current generated in the armature will be constantly passed into the brush *B'*, which will thus become the positive brush.

Theoretically the points where the brushes rest on the commutator cylinder will fall in the vertical line coinciding with the space between the poles. That is to say, the diameter of commutation, or the line connecting the points on the commutator cylinder where

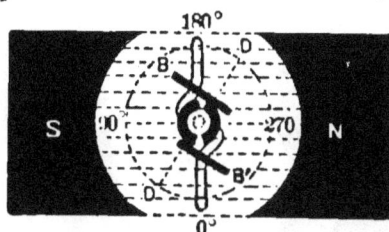

FIG. 141.—ACTION OF COMMUTATOR.

the brushes rest, will take this direction. In practice, however, this line is frequently shifted in the direction of rotation on account of the reaction which occurs between the magnetic poles of the field and the magnetic poles of the armature, as shown in Fig. 142.

Dynamo-electric machines may be divided into different classes.

(1.) According to the number and disposition of the magnetic fields into unipolar, bipolar and multipolar machines.

(2.) According to the manner in which the magnetization of the field magnets is obtained, into self-excited and separately-excited machines.

(3.) According to the character of the connections between the circuit of the magnetizing coils, the armature circuit, and the circuit external to the machine.

(4.) According to the character of the separate coils, on the field magnets, into simple and compound-wound machines.

FIG. 142.—CAUSE OF LEAD OF BRUSHES.

(5.) According to the character of the armature winding, or the shape of the armature itself.

(6.) According to whether the current developed in the armature is rendered continuous or is left alternating.

We will here discuss only those varieties of machines that arise from the different ways in which the circuit of the field magnet coils, the armature, and the external circuit are connected to one another. The most important of such varieties are :

(1.) The Series Dynamo.

(2.) The Shunt Dynamo.

(3.) The Separately-Excited Dynamo.

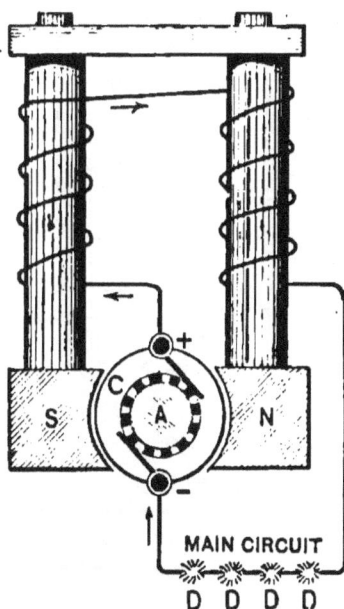

FIG. 143.—SERIES DYNAMO.

(4.) The Series-and-Separately-Excited Dynamo.

(5) The Shunt-and-Separately-Excited Dynamo.

(6.) The Series-and-Shunt-Dynamo, or the Compound Dynamo.

In the series dynamo the circuits of the field magnets and the external circuit are connected in series with the armature circuit, so that the entire armature current passes through the field magnet coils.

Such a dynamo is shown in Fig. 143.

In the series dynamo any increase in the resistance of the external circuit will decrease the power

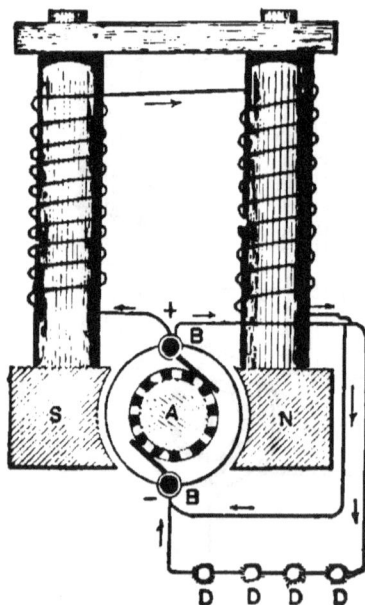

FIG. 144.—SHUNT DYNAMO.

of the machine to produce current on account of the decrease in the current of the field magnet coils.

In the same way a decrease in the resistance of the external circuit will increase the power of the machine to produce current from the increase in the mag.

netizing current. In practice these difficulties are avoided by means of automatic or hand regulators.

In the shunt dynamo, shown in Fig. 144, the field magnet coils are placed in a shunt to the external circuit, so that a portion only of the current generated passes through the field magnet coils, but all the difference of potential of the armature acts at the terminals of the field circuit.

In the shunt dynamo any increase in the resistance of the external circuit causes a smaller proportion of current to pass momentarily in the external circuit and a larger proportion to pass in the field magnet circuit, and the resulting increase in the magnetism causes an increase in the current produced in the armature. On a decrease in the external resistance, the reverse effects follow. A properly proportioned shunt dynamo will therefore be self-regulating.

When the armature of either a series or a shunt dynamo begins to rotate, the current produced in its coils, under the influence of the weak residual magnetism of the field magnets, passing through the magnetizing coils of the field magnets, increases the magnetic intensity of the machine, and, thus reacting on the armature, causes a more powerful current to flow through the field magnet coils. This again increases the strength of the magnetic field, and again reacts to increase the current strength in the armature coils, and the action continues as the machine thus, as is technically said, "builds up" until it produces its full output.

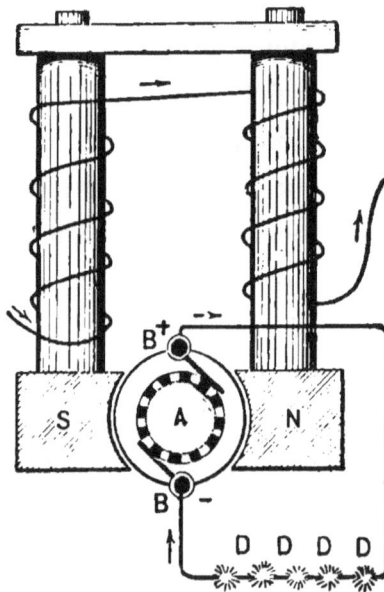

FIG. 145.—SEPARATELY-EXCITED DYNAMO.

This action is called the reaction principle of the dynamo, and was first discovered by Soren Hjorth, of Copenhagen, and afterward rediscovered independently by Siemens and Wheatstone.

In the separately-excited machine, shown in Fig. 145, the field magnet coils have no connection with the armature coils, but receive their current from a separate machine or other source.

In the series-and-separately-excited dynamo electric machine, as shown in Fig. 146, the field magnet cores are wound with two separate coils, one of which is connected in series with the armature and the ex-

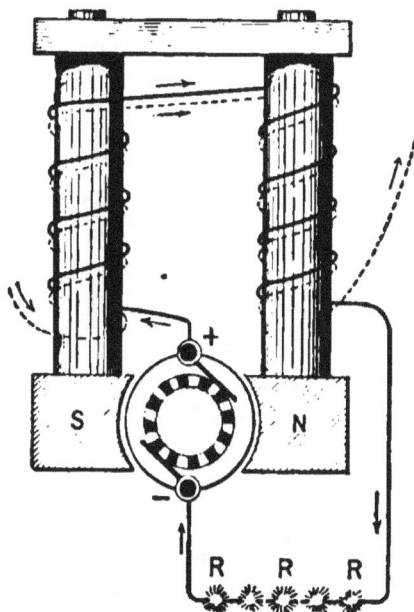

FIG. 146.—SERIES-AND-SEPARATELY-EXCITED DYNAMO.

ternal circuit, and the other with some source external to the machine by means of which it is separately excited. Since in these machines the field magnet cores have two separate and independent coils wound on them they are generally called compound-wound machines.

The shunt-and-separately-excited dynamo shown in Fig. 147 is also a compound wound machine; for, it has two independent sets of coils in the cores of its field magnets. In this type of compound-wound dynamo electric machine the field is excited both by

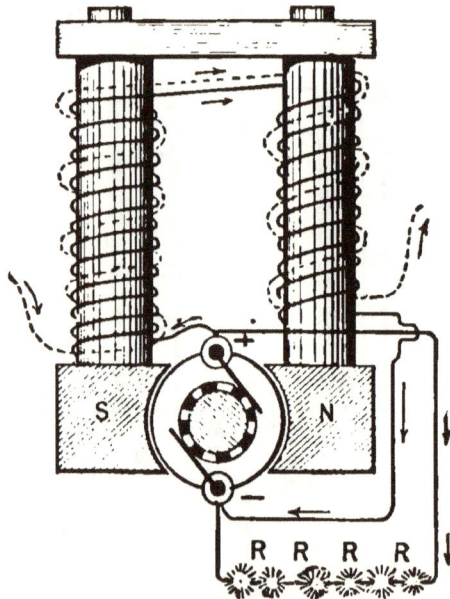

FIG. 147.—SHUNT-AND-SEPARATELY-EXCITED DYNAMO.

means of a shunt to the external circuit and by means of the current produced by a separate source.

A series-and-shunt-wound dynamo is shown in Fig. 148. In this machine the field magnet cores are wound with two separate coils, one of which is placed in series with the armature circuit and the other in shunt to the external circuit.

The series-and-shunt-wound dynamo electric machine is generally employed to maintain a constant difference of potential at its terminals. In some forms, however, the machines are over-compounded so as to increase their electro-motive force on an in-

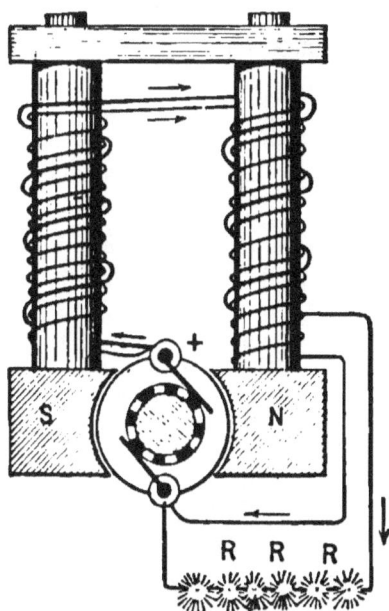

FIG. 148.—SERIES-AND-SHUNT WOUND DYNAMO.

crease of current. It is some variety of the series-and-shunt-wound dynamos that is generally used commercially. It is generally known as the compound machine, or compound-wound machine.

The great value of the use of the compound or differentially-wound dynamo electric machine, in systems of incandescent electric light distribution, in maintaining automatically a practically constant difference of potential on the mains, will be understood when it is remembered that any considerable variation in the difference of potential would render the commercial use of such lights impracticable by reason of their unsteadiness

sm

EXTRACTS FROM STANDARD WORKS.

Silvanus P. Thompson in his Third Edition of
" Dynamo-Electric Machinery,"* in the introduction
to the physical theory of the dynamo says, page 33 :

A very large number of dynamo-electric machines have
been constructed upon the foregoing principles. The variety
is, indeed, so great that classification is not altogether easy.
Some have attempted to classify dynamos according to
some constructional points, such as whether the machine
did or did not contain iron in its moving parts (which is
mere accident of manufacture, since almost all dynamos
will work, though not equally well, either with or without
iron in their armatures) ; or whether the currents generated were direct and continuous, or alternating (which is
in many cases a mere question of arrangement of parts of
the commutators or collectors) ; or what was the form of
the rotating armature (which is, again, a matter of choice
in construction, rather than of fundamental principle).

*   *   *   *   *   *   *   *   *   *   *

Suppose, then, it was determined to construct a dynamo
upon any one of these plans—say the first—a very slight

*"Dynamo-Electric Machinery: A Manual for Students of Electro-
technics," by Silvanus P. Thompson, D. Sc., B. A., F. R. S. Third
edition, enlarged and revised. London: E. & F. N. Spon, 1888.
Fourth edition. 1892. 864 pages, 498 illustrations, 29 plates. Price,
$9.00.

acquaintance with Faraday's principle and its corollaries would suggest that, to obtain powerful electric currents, the machine must be constructed upon the following guiding lines :

(*a.*) The field magnets should be as strong as possible, and their poles as near to the aimature as possible.

(*b.*) The armature should have the greatest possible length of wire upon its coils.

(*c.*) The wire of the armature coils should be as thick as possible, so as to offer little resistance.

(*d.*) A very powerful steam engine should be used to turn the armature, because,

(*e.*) The speed of rotation should be as great as possible.

Unfortunately, it is impossible to realize all these conditions at once, as they are incompatible with one another ; and, moreover, there are a great many additional conditions to be observed in the construction of a successful dynamo. We will deal with the various matters in order, beginning with the various organs or parts of the machine. Having discussed these, we take up the nature of the processes that go on in the machine when it is at work, the action of the magnetic field on the rotating armature, the reactions of the armature upon the field in which it rotates, and the various methods of exciting and governing the magnetism of the field magnets. After that we shall be in a position to enter upon the various actual types of machines for generating continuous and alternating currents.

XV.—ELECTRO-DYNAMICS.

Electro-dynamics is that branch of electricity which treats of the action of an electric current on itself, on another current, or on a magnet. The term "electro-dynamics" is employed in contradistinction to electro-statics. Electro-dynamics treats of the effects produced by electricity in motion. Electro-statics treats of the effects produced by electricity at rest.

Shortly after Oersted discovered the relation existing between electricity and magnetism, Ampère investigated the action which neighboring circuits exert on one another, when electric currents are flowing through them. After an extended series of investigations he announced the following laws:

(1.) Parallel circuits, through which electric currents are flowing in the same direction, attract each other.

(2.) Parallel circuits, through which electric currents are flowing in opposite directions, repel each other.

(3.) Circuits whose planes intersect, mutually at-

(311)

tract each other when the currents through them flow either toward or from the point of intersection.

(4.) Circuits whose planes intersect, mutually repel each other when the currents through them flow so that one approaches and the other recedes from the point of·intersection.

The correctness of these laws can be determined experimentally by a variety of apparatus.

In the apparatus shown in Fig. 149, metallic pillars

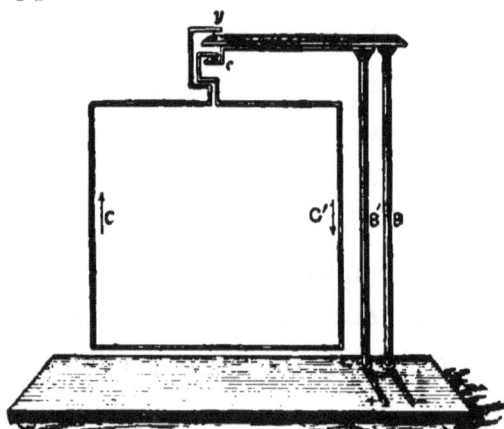

FIG. 149.—DEFLECTION OF A CIRCUIT BY A CURRENT.

B, B', support horizontal metallic arms that are provided at y and c, with mercury cups situated above one another in the same vertical line, for the insertion of the ends of a rectangular circuit C C'', of conducting wire bent at its upper extremity and provided with points extending downward. When these points are supported in mercury cups at y and

c, the rectangular circuit is suspended as shown in the figure.

In this apparatus, as in most of the apparatus employed to demonstrate Ampère's laws, a circuit is provided consisting of two parts ; namely, a fixed circuit $B B'$, and a movable circuit $C C'$. When the terminals of an electric source are connected to the points marked + and —, an electric current flows up the pillar marked B', and down the end C', of the circuit nearest thereto, returns up the left hand side of the circuit, and returns to the source down the outer pillar B.

Under these circumstances, if the plane of the movable circuit $C C'$, coincides with the plane of the fixed circuit $B B'$, when the current is caused to flow through it by the connection as shown, $C C'$, will be repelled because the current in the branch of the circuit B', is flowing in the opposite direction to the current in the branch C'.

A more convenient way of showing these movements is by means of the apparatus represented in Fig. 150. Here the fixed circuit is given the form of a coil of insulated wire $M N$, and the movable circuit $B C$, is supported at points above and below, as shown, so as to be free to move. When the connections are made as indicated in the figure, the di-

rection of the current through the two circuits is
as represented by the arrows.

Supposing, now, that before the current is passed,
the movable circuit has been placed in the same
plane as that of the coil *M N;* then, since the branch
M, of the fixed circuit, has a current flowing through
it in the opposite direction to the branch *B,* of the
movable circuit, as soon as such current begins to flow

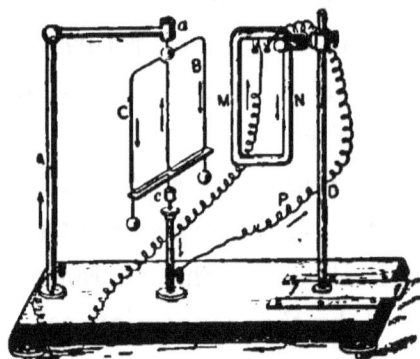

FIG. 150.—AMPÈRE'S STAND.

a repulsion occurs, and the movable circuit tends to
place itself with its plane at right angles to that of
the fixed circuit, and will so place itself, provided
the current is sufficiently strong. If, however, the
direction of the current in the branch *M,* of the fixed
circuit is the same as that in the branch *B,* of the
movable circuit (which can be effected by reversing
the position of the fixed circuit *M N*), on the passage

of the current, if the movable coil has been previously placed so that its plane does not coincide with the plane of the fixed coil, it will be attracted thereto.

In order to demonstrate the third law, the apparatus shown in Fig. 151 may be.employed. Here the movable circuit *A B C D*, supported as shown, and having a current flowing through it in the

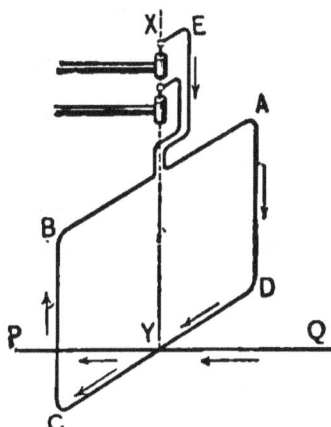

FIG. 151.—ATTRACTION OF ANGULAR CURRENTS.

direction indicated by the arrows, will be attracted by the fixed circuit *Q P*, when the current in it flows in the direction from *Q* to *P*, for then the currents in both the fixed and movable circuits are flowing in the same direction both toward and from *Y*, the point of intersection of the circuits.

If, however, the current flows through the fixed

circuit in the opposite direction, or from P to Q, then since the currents are flowing in opposite directions in the fixed and movable circuit toward and from the point of intersection Y, the fixed circuit will repel the movable circuit.

The fact that parallel currents flowing in the same direction attract each other can be shown by the following experiment : When an electric current flows through a flexible conductor wound in the shape of a loose or open spiral, the current flows in the same direction through the contiguous turns of the spiral and the mutual attractions exerted between such turns cause them to attract one another, thus shortening the spiral. If the spiral conductor is supported at its upper end, so that its lower end dips into a mercury surface, and the current flowing through the spiral is made sufficiently strong, the mutual attractions will shorten the spiral sufficiently to cause it to lift itself out of the mercury surface and thus break the circuit. The spiral then falls and again makes contact with the mercury surface and causes the current to again flow through it. There is thus established a succession of automatic makes-and-breaks of the circuit that follow one another with considerable rapidity. A brilliant spark, caused by the induced current on breaking the circuit, appears each time the spiral leaves the surface of the mercury.

Electro-dynamic attractions and repulsions are also produced by the action of magnets on movable circuits. This should be expected, since electric circuits possess all the properties of magnets.

If the magnet *A B*, be placed parallel to the movable circuit shown in Fig. 152, in the direction

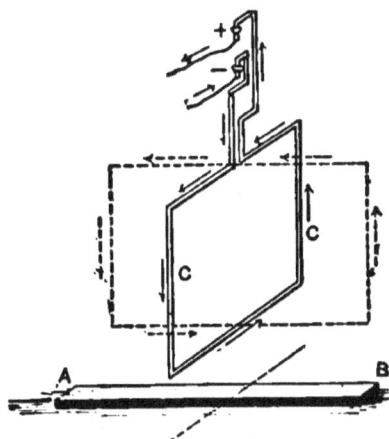

FIG. 152.—DEFLECTION OF CIRCUIT BY A MAGNET.

shown by the dotted lines, and the current be then passed through the movable circuit *C C*, in the direction indicated by the arrows, the circuit will move and tend to place itself at right angles to the axis of the magnet, or in the position shewn by the full lines in the figure. A careful study of these movements will show that they are the same as would occur if an electric current were circulating around the magnet in the same direction as the currents which Ampère assumed to exist in all magnets and to be the cause of their magnetism.

It is a well-known fact that when a current flows through the coils of a solenoid, that is, of a cylindrical coil of wire the convolutions of which are circular, the solenoid acquires all the properties of a magnet, so that it will be attracted or repelled by another magnet, or by another solenoid placed near it, through whose circuit an electric current is flowing.

As in the case of magnetic attractions and repulsions, like solenoidal poles repel and unlike poles attract. This action of solenoids on one another will be understood from a consideration of the directions of the currents required to produce the respective north or south poles.

Unlike poles of a solenoid attract each other because the currents which produce such poles flow in the same direction in parts of the circuits which lie nearest to each other.

In the same manner like poles of a solenoid repel each other, because the currents which produce such poles flow in the opposite direction, in parts of the circuit which lie nearest each other.

The cause of the attractions which exist between

the unlike poles of solenoids is to be found in the direction of the lines of magnetic force which are produced by the currents flowing through such solenoids.

Like charges of electricity and like magnet poles repel, and there is, therefore, to many students a difficulty in understanding why parallel currents flowing in the same direction in neighboring circuits should attract each other, but the reason is apparent

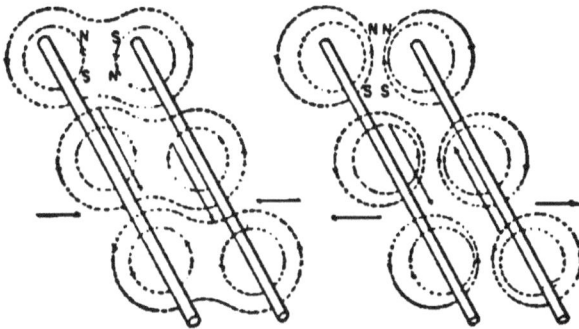

FIG. 153.—MUTUAL ACTION OF MAGNETIC FIELDS.

when traced to the magnetic fields which such currents produce.

The cause of this will be understood from a study of Fig. 153, which shows on the right hand two circuits extending parallel to each other through which currents are flowing in opposite directions, and on the left hand side two parallel circuits through which currents are flowing in the same direction. The small arrows show the directions of the lines of magnetic force produced by the current.

Tracing the direction of the circular lines of magnetic force, which are produced by the currents flowing through the circuit on the left, it will be noticed that their lines of magnetic force in those parts of the circuit where the lines lie nearest to one another extend in opposite direction.

Parallel circuits, therefore, flowing in the same direction, attract one another because their approached lines of magnetic force extend in opposite directions, and oppositely directed lines of magnetic force attract one another.

Similarly, if the drawing at the right hand be inspected, which represents the magnetic fields produced by two parallel circuits, the currents through each of which are flowing in opposite directions, it will be seen that their lines of force have the same direction in parts of their circuits which lie nearest together, and that these lines of force extending in the same direction repel one another.

Parallel currents, therefore, flowing in opposite directions repel one another, because their approached lines of magnetic force extend in the same direction.

The mutual action, therefore, of parallel currents

is thus to be traced to the action of the magnetic lines of force produced by the currents.

Generally these laws may be expressed as follows : Lines of magnetic force extending in opposite directions attract one another ; lines of magnetic force extending in the same direction repel one another.

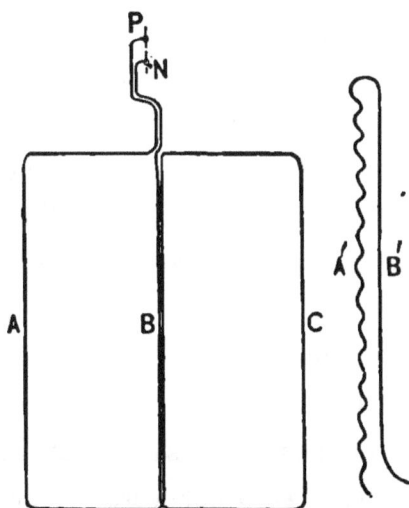

FIG. 151.—RECTILINEAR EQUIVALENT OF SINUOUS CURRENT.

Ampère proved that when a circuit is bent on itself, so that the current flows in one part of the circuit in the opposite direction to that in which it flows in the remainder of the circuit, the two parts exert no force of magnetic attractions or repulsions on external objects. This expedient of doubling a wire on itself is adopted in the manufacture of resistance coils, in order to prevent the magnetic fields produced by such coils exerting any disturbance on the needle of the galvanometer.

If a circuit is bent, as shown to the right in Fig. 154, so that one part of it, as at B', is formed of a straight conductor and the other portion, as A', is formed of a zig-zag conductor, although the portion A', is longer than the portion B', yet if a current be sent into such a wire at one end and passed out at the other end it will produce no action on the movable circuit A B C, when approached to it, because the current flowing through the branch B', neutralizes the effect of the current flowing through the branch A'.

The term sinuous current is sometimes applied to a current flowing through a sinuous conductor.

Successive portions of the same rectilinear current repel one another. In other words, a current flowing through one part of a straight or rectilinear conductor tends to repel the part lying nearest to it and is itself repelled.

A circuit O A, Fig. 155, movable around O, as a centre, will be continuously rotated in the direction shown by the curved arrow by the current flowing through the rectilinear circuit Q P, in the direction shown. If the direction of the current through the movable circuit be as indicated by the smaller arrows, there will be attraction in the positions corresponding to (1) and (2), and repulsion in the position corresponding to (4).

A conductor through which a current of electricity is flowing tends to rotate continuously around a magnet pole, as can be shown by suitably mounting a conductor so as to be capable of rotation around a magnet. When a current of electricity is sent through such a conductor, the conductor will ro-

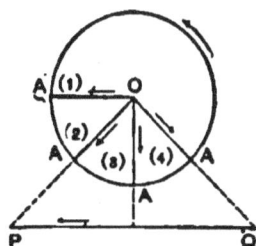

FIG. 155.—CONTINUOUS ROTATION OF CURRENT.

tate continuously in one direction around the north magnetic pole, and in the opposite direction around the south magnetic pole.

By reason of the mutual actions exerted between a conductor through which a current of electricity is flowing and a magnet pole the circuit tends to wrap or twist itself around such pole, as can be shown by the following experiment suggested by Lodge :

If a powerful current of electricity is passed

through some eight feet in length of gold thread, such as is employed for making gold lace, hung in a vertical position near a vertical bar magnet, as soon as the current passes through the thread it will wrap itself around the bar magnet, one-half twisting itself around the north pole, the other half around the south pole.

Or, the same thing can be shown by the following experiment suggested by S. P. Thompson:

A stream of mercury, which is falling between the poles of a powerful electro-magnet, will, when an electric current is flowing through the coils of the magnet so as to energize it, be twisted in a spiral direction, which will vary both with the direction of the current or with the magnetic polarity.

EXTRACTS FROM STANDARD WORKS.

In the revised edition of the "Student's Text-Book of Electricity,"* by Noad, on page 264, the following statements are made concerning the action of circuits through which electric currents are passing on magnets:

The grand fundamental fact observed by Oersted in 1819 was that when a magnetic needle is brought near the connecting medium (whether a metallic wire or charcoal, or even saline fluids) of a closed voltaic circuit, it is immediately deflected from its position, and made to take up a new one, depending on the relative positions of the needle and conductor.

*   *   *   *   *   *   *   *   *

The extent of the declination of the needle depends entirely on the *quantity* of electricity passing along the conductor; it has nothing to do with its *tension*, which is probably the reason that the first inquirers failed to discover the above effects since they all worked with statical electricity.

When the current is rectilinear, the length of the conducting wire considerable, so that in relation to that of the needle it may be regarded as infinite, the intensity of the electro-magnetic force was shown by Biot and Savary to be "*in the inverse ratio to the simple distance of the magnetized*

*"The Student's Text-Book of Electricity," by Henry M. Noad, Ph. D , F.R.S. Revised by W. H. Preece. M.I.C.E. London: Crosby Lockwood & Co. 1879. 615 pages, 471 illustrations. Price $4.00.

needle from the current;" but it is only under these conditions that the law is true, for it has been shown by Laplace that the elementary magnetic force—that is, the elementary action of a simple section of the current upon the needle—is, like all other known forces, in the inverse ratio of the *square of the distance,* and proportional to the *sine of the angle* formed by the direction of the current, and by the line drawn through the centre of the section to the centre of the needle. In fact, by calculating, according to this principle the sum of all the elementary actions that are exercised on a small needle by an indefinite rectilinear current, it is found that the intensity of this resultant should be, as experiment proves it really is, in the inverse simple ratio of the distance.

XVI.—THE ELECTRIC MOTOR.

An electric motor consists of any combination of parts by means of which electrical energy is converted into mechanical energy.

In electric motors, as now generally constructed, the electrical energy is caused to produce mechanical energy by means of the attractions and repulsions which are exerted between the magnetic fields of electro-magnets, or between the magnetic fields of electro-magnets and the magnetic fields produced by currents flowing through neighboring conductors.

An electro-magnetic motor depends for its operation on the tendency of a conductor through which a current of electricity is flowing to move in a magnetic field in accordance with the principles of electro-dynamics already pointed out. Such a tendency to motion arises from the magnetic attractions and repulsions which the two fields exert on each other. The entire magnetism may be produced by a current, or part by a current, and the rest by the field of a permanent magnet. In actual practice, however, electro-magnets are generally employed.

One of the simplest and earliest forms of electric motors was that in which the motion was obtained from the interactions existing between the magnetic field of a conductor through which a current of electricity is passing, and that produced by the poles of a permanent magnet. Such a form of apparatus, which was known as Barlow's wheel, is shown in Fig. 156. Here a metallic wheel rotates in the magnetic field produced by the poles of a permanent horseshoe magnet, in the direction shown by the

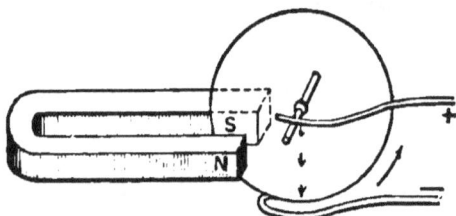

Fig. 156.—Barlow's Wheel.

curved arrow, when a current of electricity is sent through it from the axis to the circumference.

A dynamo-electric machine is capable of acting as a motor if a current of electricity is sent through its circuit. In point of fact the construction of the modern electric motor in general is based on the same principles as the construction of dynamo-electric machines.

The discovery of the reversibility of a dynamo-electric machine, or its ability to operate as a motor when a current of electricity is passed through it may be said to have been the discovery on which the introduction of the electric motor, as it exists at the present day, is based.

The following analogies can be shown to exist between dynamo-electric machines and electric motors:

(1.) If mechanical energy is applied to a dynamo-electric machine, so that its armature is caused to rotate, a difference of electromotive force is produced in such armature; and, conversely, if electric energy be applied to a dynamo-electric machine, by sending a current through its circuit, the armature will rotate and produce mechanical energy.

(2.) If mechanical energy is applied to a properly designed dynamo-electric machine such, for example, as that made for incandescent lighting, so that its armature is run at a constant speed, the electromotive force which it produces will remain practically constant no matter what load may be on the machine, or how much current it is generating; conversely, if electric energy is applied to an electric motor, designed so as to have a constant field, by maintaining a constant difference of potential at its terminals, it will run at a constant speed no matter what load may be placed on it.

In an electric motor the pull produced along the

circumference of the shaft by the electro-dynamic action of the fields, or, in other words, the amount of turning force which such shaft exerts, is called its torque. In a well-constructed motor, as the load on the motor increases, the torque increases proportionally.

By the efficiency of an electric motor is meant the ratio which exists between the electric energy required to drive the motor and the mechanical energy which it gives out.

The efficiency of an electric motor may be made very high; it may even rise to not far from 100 per cent. When, however, a motor is overloaded its efficiency rapidly decreases.

When mechanical power is applied to drive the armature of a dynamo, the circuit of which is closed, an electric current is produced which flows through the circuit of the machine and sets up in its field magnets a magnetic polarity, which will be of such a character, as compared with that produced by its armature, that the armature will oppose or resist being moved in the direction in which it is moved in order to produce differences of potential.

If, however, it is compelled to move in this direction past the field magnet poles, the mechanical energy so expended is, in accordance with the well-

known principle of the conservation of energy, converted into a very nearly equal amount of electrical
energy which appears in the current flowing through
the circuit connected therewith.

Suppose, for example, that the polarity produced
in the armature is of such a character, as compared
with that produced in the field magnets, that the
poles n and s, are produced in those parts of the armature core that lie in the vertical gap between the
poles N and S, of the field magnet, as shown in Fig.
157.

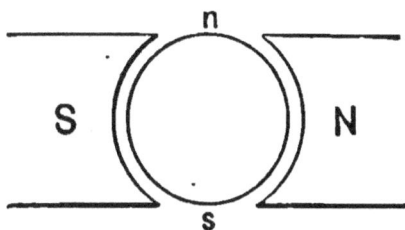

FIG. 157.—POLARITY OF ARMATURE AND FIELD.

If the armature is moved in a direction such that
the n, pole at the top of the armature core moves
toward the s, pole of the field magnets, that is, if the
top of the armature is moved toward the left hand,
no energy will be expended against the magnetic field
and no current will be produced.   If, however, the
armature is driven so that the top moves from left
to right, then the north pole at the top of the armature is moved toward the north pole of the field

magnet, and the south pole at the bottom is moved toward the south pole of the field magnet, and the energy so expended is converted into electrical energy.

If, now, such an armature be supplied with an electric current from some source outside the machine, and the current flows through the field magnets and the armature in such a direction that the polarity remains the same as shown in Fig. 156, it can readily be seen that the armature will turn in the opposite direction to that in which it requires to be turned by power in order to produce differences of potential; for now the north pole at the top of the armature will be attracted and will move toward the south pole of the field magnets, and the south pole at the bottom of the armature will be attracted and will move toward the north pole of the field magnets.

A difference exists between the position the collecting brushes, or, more properly speaking, the distributing brushes, must have on the commutator cylinder of an electric motor and on the commutator cylinder of a dynamo-electric machine in order that there shall be no excessive sparking. In both cases a lead must be given to the brushes on account of the reaction which exists between the fields produced by the field magnets and that produced by the armature core. In the case of a motor, according to S. P. Thompson, the direction of this lead will be forward, or in the direction of rotation, if it is only desired to obtain a rapid rotation. But, if it is desired to obtain the position of least-sparking it must be moved a short distance in the opposite direction. In other words, the lead which must be given to the brushes on the commutator of a motor is in the opposite direction to that which must be given to the brushes on the commutator of a dynamo.

For the purpose of avoiding too great a lead for distributing brushes, the magnetism generated in the field magnets should be made great as compared with that generated by the armature. It may be observed in this connection, that with the best construction of motors the necessity for any lead for the brushes becomes very small.

In actual practice it is often necessary to be able to readily change the direction in which the motor is running. Such a change in direction can be obtained in the following ways :

(1.) By reversing the direction of the current through the armature.

(2.) By reversing the direction of the current through the field magnets, but not by reversing the direction of the current through the armature and the field magnets simultaneously.

The reversing of the direction of the current can be effected by changing the position of the distributing brushes, and at the same time changing the position of the lead, provided the motor works under a sensible lead.

A little consideration will show that in an ordinary bi-polar dynamo-electric machine, by rotating the brushes through 180°, less an amount equal to twice the angle of the ordinary lead (because the lead must then be in the opposite direction to what it formerly was), will reverse the direction of the motor's motion.

Any device, therefore, by which the brushes can be readily moved through such a distance will effect a reversal of the motion of the motor.

This method of reversal has been employed by Reckenzaun and others. Reckenzaun's reversing gear is shown in Fig. 158. In it there are two pairs of brushes, each pair of which is fixed to the common brush-holder and is capable of turning on a pivot and of being moved or rotated by the motion of a lever connected therewith.

We will now consider some peculiarities concerning the direction of the motion that will be produced

when an electric current is sent through the circuit
of dynamo-electric machines of different types.

A series dynamo will operate as a motor when a
current is sent through it in a direction opposite to
that of the current which it produces when in operation as a generator; the polarity of the field is re-

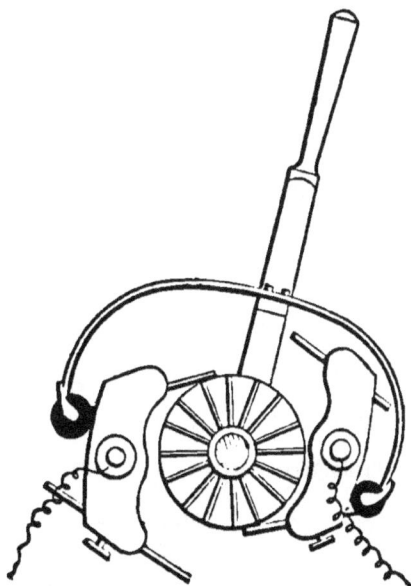

FIG. 158.—REVERSING GEAR FOR ELECTRIC MOTORS.

versed and the dynamo will turn as a motor in the
opposite direction to that required to produce the
current. If the current is reversed, the polarity of
both the field and the armature will be reversed, and
the machine will still rotate as a motor in the oppo-

site direction to that in which it rotated as a generator.

A series dynamo, therefore, always rotates as a motor in a direction opposite to that in which it is rotated as a generator, unless the polarity of its field magnets or its armature only is reversed, when it may be made to rotate in the same direction as it is rotated as a generator.

A shunt dynamo when operated as a motor will turn in the same direction as that in which it is turned as a generator; for, if the direction of the current in the armature is the same as in the generator, that in the shunt is reversed.

A separately-excited dynamo will operate as a motor when a current is sent through its armature, and will always turn in a direction opposite to that in which its armature requires to be turned in order to produce a current in the same direction as the driving current.

A compound-wound dynamo when operated as a motor will move in a direction opposite to that of its motion as a generator when its series coils are more powerful than its shunt coils, and in the same direction when its shunt coils are more powerful than its series coils.

If a galvanometer is placed in the circuit of a

motor, the terminals of which are maintained at a constant difference of potential, and the armature of the motor is fixed so as to be unable to move, a certain current will flow through the circuit of the motor as may be determined by the deflection of the galvanometer needle. If, now, the armature of the motor be permitted to move, it will be found that the more rapid its motion becomes the smaller will be the current that passes through the circuit, as will be seen by a smaller deflection of the galvanometer needle.

The cause of this is as follows : As the armature moves through the field of the machine, its coils of wire cut the lines of magnetic force and, just as in a dynamo, will have differences of potential generated in them, which are opposed to the difference of potential of the current which drives the motor. In this way a counter-electromotive force is set up, which acts like a resistance to oppose the passage of the driving current through the coils of the motor. Therefore, as the speed of the motor increases, the strength of the driving current becomes less, until, when the maximum speed is reached, very little current passes.

When, however, a load is placed on a motor, so that it is caused to do work, its speed tends to decrease, and the counter-electromotive force is decreased, thus permitting a greater driving current to pass through 'the circuit.   In this way a motor automatically regulates the current required to drive it.   For this reason, therefore, electric motors are very economical in operation, provided they are efficient at full load.

The relations between the power required to drive the generating dynamo and that produced by the electric motor, through which its current passes, are such that the maximum work per second is done by the motor when it runs at such a rate that the counter-electromotive force it produces is half that of the current supplied to it.   The maximum rate of work of an electric motor is therefore done when its theoretical efficiency is only 50 per cent.   This, however, must be carefully distinguished from the maximum efficiency of an electric motor.   A maximum efficiency of 100 per cent. can be attained theoretically, and considerably over 90 per cent. is attained in practice.   In such cases, however, the motor is doing work at less than its maximum rate.

An efficiency of 100 per cent. would be theoretically reached when the counter-electromotive force of the motor is equal to that of the source supplying the driving current.   If the driving machine is of the same size and type as the motor, the two machines would be running at the same speed. If, now, a load is put on the motor so as to reduce its speed, and thus prevent it from producing a counter-electromotive force of more than 90 pei cent., its efficiency will be about 90 per cent. In such a case, therefore, the efficiency is represented by the relative speeds at which the generator and motor are running.

Jacobi's law of maximum effect, namely, that an electric motor gives its maximum work when it is geared to run at a speed which reduces the current to half the strength it would have when at rest, does not apply to motors in practice on account of limitation of current carrying capacity. For example, a motor of nine horse power and 90 per cent. efficiency loses one horse power in heating itself, and would be injured if much more than one horse power were converted into heat in it. If run according to Jacobi's law half of a greater amount than nine horse power would be converted into heat in itself, and this would overheat it.

If the current from an alternating current dynamo is sent through a similar alternating current dynamo brought up to the same speed, it will drive it as a motor. Most alternating current motors, however, possess the disadvantage of requiring to be maintained at exactly the same speed as that of the driving dynamo, with the additional disadvantage that they require to be brought up to exactly this speed before the driving current can be supplied to them; overloading which reduces the speed, even by the smallest amount, will therefore stop the motor. Considerable improvements, however, are being intro-

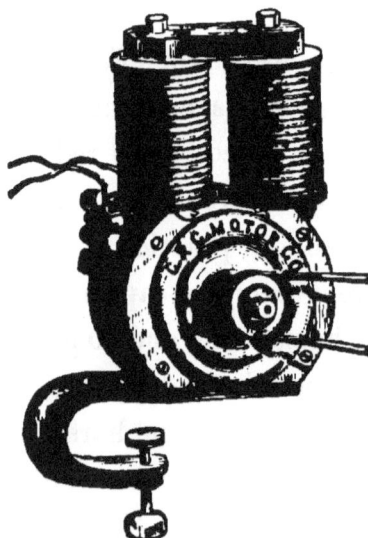

FIG. 159.—THE C. &. C. MOTOR.

duced into alternating current motors, by which these difficulties are almost entirely removed.

In Fig. 159 is shown a form of electric motor suitable for small work, called the C. & C. motor, from the initials of its inventors, Curtis and Crocker. Its armature, which consists of a ring-wound core, is completely enclosed so as to protect it from injury, either from dust or other causes.

In a compound-wound motor the field magnets have two windings which oppose each other, so that the speed remains constant no matter what may be

Fig. 160.—The Sprague Electric Motor.

the load. In compound-wound motors the series coils are wound differentially to the shunt coil, so that one tends to demagnetize the field magnets, while the other tends to magnetize them.

A form of compound-wound motor is shown in Fig. 160. It is one of the forms of the Sprague motor.

In a curious form of motor, known as the pyro-magnetic motor, the motion is obtained by the at-

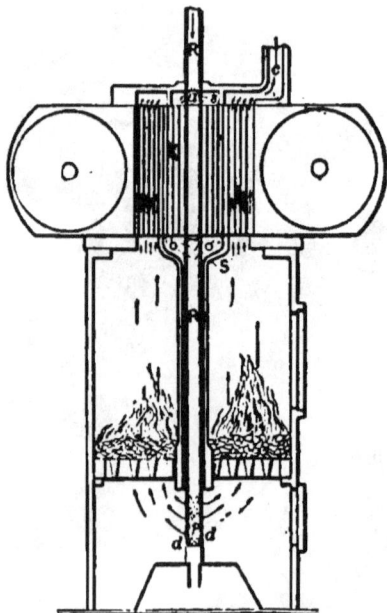

FIG 161.—EDISON PYRO-MAGNETIC MOTOR.

traction which magnet poles exert on an unequally heated movable disc of iron.

The intensity of magnetization of iron decreases with an increase of temperature, the iron losing all its magnetism at a red heat. If, therefore, a disc of iron so placed between the poles of a magnet as to be capable of rotation, is heated at a part which lies nearer one pole than the other, it will be caused to rotate, since it becomes less powerfully magnetized at the heated part.

Such a form of motor does not at present possess very great efficiency.

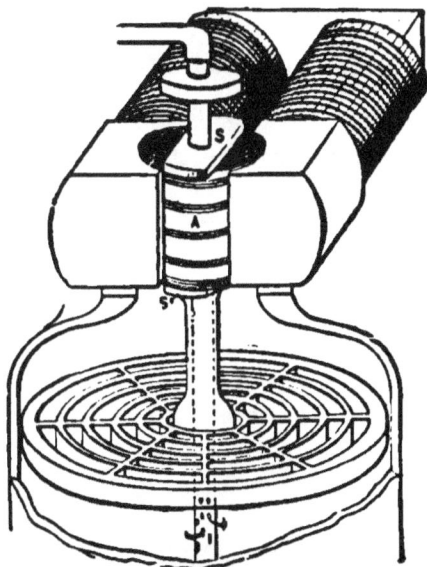

FIG. 162.—EDISON PYRO-MAGNETIC MOTOR.

A form of pyro-magnetic motor devised by Edison is shown in Fig. 161 in vertical section and in Fig. 162 in elevation.

A movable cylinder of iron *A*, formed by a bunch of small iron tubes, is heated by the products of combustion of a fire placed beneath them. To render this heating local, a flat screen *S*, is placed dis-symmetrically across the top to prevent the passage of hot air through the portion of the iron tubes so screened. The air is supplied to the furnace by passing down from above through the tubes so screened and thereby cools them. This is shown in the drawings, the direction of the heated and cooling air currents being indicated by the arrows. Supplying the air from above insures a more rapid cooling of the screened portion of the tubes.

EXTRACTS FROM STANDARD WORKS.

Concerning the reversibility of the dynamo-electric machine, Martin and Wetzler, in " The Electric Motor and Its Applications,"* on page 29, speak as follows :

But there is another version of the story, told by M. Hippolyte Fontaine to the Société des Anciens Elèves des Ecoles Nationales des Arts et Métiers. M. Fontaine claims to have actually invented or discovered the electrical transmission of power, as will be seen from the following short extract from his paper, read before the above-mentioned society:

On the 1st of May, 1873—that is, on the date fixed four years previously by imperial decree—the Exhibition in Vienna was formally opened. At that time the machinery hall was as yet incomplete, and remained closed to the public until the 3d of June, when it was also thrown open. I was then engaged with the arrangement of a series of exhibits, shown for the first time in public, which were intended to work together, or separately, as desired. There was a dynamo machine by Gramme for electroplating, giving a current of 400 ampères at 25 volts, and a magneto machine, which I intended to work as a motor from a primary battery, or from a Planté accumulator, to demon-

*"The Electric Motor and Its Applications," by T. C. Martin and Jos. Wetzler. With an appendix on The Development of the Electric Motor since 1888, by Dr. Louis Bell. New York: The W. J. Johnston Company, Ltd. 1893. 325 pages, 354 illustrations. Price, $3.00.

strate the reversibility of the Gramme dynamo. There was also a steam engine of my invention heated by coke, a domestic motor of the same type heated by gas, a centrifugal pump placed on a large reservoir, and arranged to feed an artificial cascade, and numerous other exhibits. To vary the experiments I proposed to show, I had arranged the pump in such a way that it could be worked either by the Gramme magneto machine or by the steam engines (Fontaine).

On the 1st of June it was announced that the machinery hall would be formally opened by the Emperor at 10 A. M. on the day after the morrow. Nothing was then in readiness, but those who have been in similar situations know how much can be got into order in the space of 48 hours just before the opening of a great exhibition. In every department members of the staff with an army of workmen under their orders were busily clearing away packing boxes and decorating the space allotted to the different nations. These gentlemen visited all the exhibits in order to determine which of them should be selected for the special notice of the Emperor, so as to detain him as long as possible among the exhibits of their respective countries.

M. Roullex-Duggage, who superintended the work in the French section, asked me to set in motion all the machinery on my stand, and especially the two Gramme machines. I set about at once, and on the 2d of June I had the satisfaction of getting the large Gramme dynamo, the two engines (Fontaine), and the centrifugal pump to work but I failed to get the motor into action from the primary or secondary battery. This was a great disappointment, especially as it prevented my showing the reversibility of the Gramme machine. I was puzzled the whole of the evening and the whole of the night to find a means to accomplish my object, and it was only in the morning of June 3, a few hours before the visit of the Emperor, that the idea struck me to work the small machine by means of a derived circuit from the large machine. Since I had no leads for that purpose, I applied to the representatives of Messrs. Manhis, of Lyons, who were kind enough to lend me 250 metres of cable, and when I saw that the magneto machine was not only set in motion, but developed so much power as to throw the water from the pump beyond the reservoir, I added more cable until the flow of water became normal. The total length of cable in circuit was then over two kilometres. This great length gave the idea that by the employment of two Gramme machines it would be possible to transmit mechanical energy to great distances. I spoke of this idea to various people, and I published it in the *Revue Industrielle* in 1873, and subsequently in my book on the Vienna Exhibition. The publicity thus given to it was so great that I had neither time nor desire to protect my invention by a patent. I must also mention that M. Gramme has told me that he had already worked one dynamo by the other, and I have always held that the honor of my experiment belongs to the Gramme Company.

In the fourth edition of his "Dynamo-Electric Machinery,"* S. P. Thompson, on page 548, speaks thus of the actions existing between a magnet and a conductor through which a current of electricity is flowing:

In the first chapter, the definition was laid down that dynamo-electric machinery meant "machinery for converting energy in the form of mechanical power into energy in the form of electric currents, or *vice versa.*" Up to the present point we have treated the dynamo solely in its functions as a generator of electric currents. We now come to the converse function of the dynamo, namely, that of converting the energy of electric currents into the energy of mechanical motion.

An electric motor, or, as it was formerly called, an electro-magnetic engine, is one that does mechanical work at the expense of electric energy ; and this is true, no matter whether the magnets which form the fixed part of the machine be permanent magnets of steel or electro-magnets. In fact, any kind of dynamo, independently excited or self-exciting, can be used conversely as a motor, though, as we shall see, some more appropriately than others. But, whether the field magnets be of permanently magnetized steel, or of temporarily magnetized iron, all these motors are electro-magnetic in principle ; that is to say, there is some part either fixed or moving which is an electro-magnet, and which as such attracts and is attracted magnetically.

---

* "Dynamo-Electric Machinery : A Manual for Students of Electrotechnics," by S. P. Thompson, D. Sc., B. A. Fourth Edition. Enlarged and Revised. London: E. & F. N. Spon. 1892. 864 pages, 498 illustrations, 29 plates. Price, $9.

XVII. — ELECTRIC TRANSMISSION OF POWER.

Any system for the electric transmission of power consists essentially of the following parts :

(1.) Of a line conductor or circuit established between the two stations.

(2.) Of an electric source, or battery of electric sources, at one of the stations, generally in the form of a dynamo-electric machine, or battery of dynamo-electric machines, for the purpose of converting mechanical energy into electrical energy.

(3.) Of various electro-receptive devices placed in the circuit of the line wire or conductor, in the form of electric motors, for the purpose of reconverting electrical energy into mechanical energy.

The electro-receptive devices may be connected to the line wire or conductor either in series or in multiple ; or, in other words, the circuits established between the two stations may be either constant-current circuits or constant-potential circuits.

Strictly speaking, all electric circuits are established for the transmission of electric energy ; for, in all circuits, some form of energy is expended at the source for the purpose of producing electric energy, which is transmitted over a line wire or conductor connecting such source with an electro-receptive device or devices, in which such energy is utilized.

A system for the electric transmission of power differs from the above merely in the fact that in such a system mechanical power is transmitted through considerable distances between the place where it is generated and the place where it is utilized. Some prime mover, as, for example, a steam engine converts the energy of heat into mechanical energy, and which mechanical energy is converted by means of a dynamo-electric machine into electrical energy, is employed at one end of the line, and the electrical energy so obtained is converted by means of such electro-receptive devices as electric motors into mechanical energy at the other end of the line ; or, as is very frequently the case, the mechanical energy of a water power is converted at one end of the line by means of a dynamo into electric energy, which is employed as before.

The electric transmission of power possesses marked advantages over any other known system for the transmission of power, such, for example,

as by means of belting, wire ropes, gears, or by
means of compressed air or other fluids.

Among the advantages possessed by the electric
transmission of· power the following may be mentioned :

(1.) The distance through which power may be
economically transmitted electrically is very much
greater than by any other known means. In the
case of belting or wire rope transmission the limits
of such transmission are measured in feet, while
in the case of electric transmission they are measured in miles. Hydraulic transmission, though
very economical within certain limits of distance,
cannot compete for great distances with electric.

(2.) Sources of power can be utilized by systems
of electric transmission that would otherwise be impracticable. Suppose, for example, a waterfall is
situated at a fairly considerable distance from the city
or other location where it is desired to establish a
manufacturing plant. A water-wheel can be placed
at such waterfall for the purpose of converting
mechanical energy by means of a dynamo into electric energy, and the electric energy so developed can
be led by means of conducting wires to the distant
place, where, on being passed through electric motors,
it can be converted into mechanical power.

(3.) The means by which the driving power is connected to the distant driven mechanism is far simpler than by any other means, a mere pair of wires or conductors, which may pass in any direction, and around any number of corners or bends, being all that is necessary for this purpose.

Contrast this simplicity of detail with belts or wire ropes, or even with compressed air or with water, and the advantages will be self-evident ; the difficulties of leakage at joints and the cost of construction in the case of the transmission by compressed air being very much greater than in the case of electric transmission.

The utilization of water power for the production of electric power, and the transmission of such power to great distances, is rapidly coming into extended use. Numerous cases exist in which such transmission is actually being carried on over very considerable distances. Even such rivers as Niagara near the Falls are about to be made to expend a portion of their energy in performing useful work, and there is every prospect, in the near future, that the number of such practical applications will increase.

The percentage of the relative efficiencies of the steam engine and the electric motor is very much in favor of the electric motor. The efficiency of the best steam engine is only in the neighborhood of 17 per cent., while that of the electric motor can be made almost as high as required, it often exceeding 95 per cent. The question then arises, Can we ever expect the electric motor to replace the steam engine? The answer would appear to be that such displacement must certainly occur whenever electricity can be produced more economically by the burning of coal, or by means other than through the intervention of the steam engine.

As long as the best method of producing electrical energy is limited to driving a dynamo by a steam engine the steam engine must of course hold its own. If, however, the discovery is ever made—and such discovery is by no means improbable—of a means of economically producing electricity directly from the burning of coal, then the steam engine will certainly be replaced by the electric motor.

Even during the present day the electric motor can economically compete with steam under the following circumstances:

(1.) In certain cases where it replaces horse or other animal power.

The cost of producing and distributing electric energy for replacing the power of horses on street railways is much less than feeding and caring for the horses.

(2.) For all cases where an available water power exists, even though at considerable distances from the places where its energy is to be utilized.

The energy of moving water, except in the case of very irregular streams, can generally be caused to produce mechanical power at a smaller expense than by the steam engine.

(3.) In places where only small amounts of power are required.

A saving can generally be made, when but a comparatively small power is needed, by using electric power under circumstances where a steam engine and boiler and the services of an engineer can thereby be dispensed with. In manufacturing centres where rentals are high the mere saving of the space required by a steam plant secures so great economy as to permit the displacement of steam power by electric power. This is now done economically in a number of cases either by the establishing of a large central power station driven by water, or by steam power, and supplying outlying districts with electric power.

The various systems of telegraphic and telephonic communication afford excellent illustrations of the

actual transmission of electric power. In the telephone the voice of the speaker, acting as the driving
power, converts mechanical energy into electrical
energy, which is transmitted over a line wire or conductor, and, passing at the distant end through a
form of electro-receptive device called a receiver,
converts the electrical energy into mechanical energy,
which in its turn reproduces the articulate speech
spoken into the transmitting instrument.

FIG. 163.—TELPHERAGE SYSTEM.

The electric transmission of power is not limited
to the case where electric energy is sent into a line
wire or conductor at one end, and is utilized at the
other end only ; for, in many cases, the energy is
taken from such line wire or conductor at intermediate points as well as at its further end.

Examples of this are seen in various telpherage

systems, in the porte-electric system, and, especially, in any of the well-known systems of electric railways.

The telpherage system, an invention of Fleeming Jenkin, is a system whereby carriages suspended from electric conductors are propelled or driven along said conductors by the action of electric motors that take the current required to energize them from the conductors over which they move.

The telpherage system shown in Fig. 163 is called the cross-over or parallel system. In this system two conductors, connected to a dynamo-electric machine,

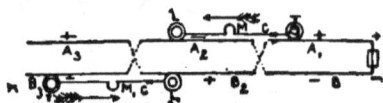

FIG. 164.—CIRCUIT FOR TELPHERAGE SYSTEM.

which maintains them at a constant difference of potential, are caused to cross each other at regular intervals in the manner shown in Fig. 164. By this means the wires or conductors on each side of the road are alternately positive and negative, and two lines are thus provided—an up and a down line.

The train of cars as shown at $L\ T$, or at $L'\ T$, is of sufficient length to make contact with two adjoining sections at the same time. The wheels of each train are insulated from their trucks, but are connected together in pairs by means of a conductor. Therefore, as the train passes over the track a current flows through the motor on each train from the positive to the negative section.

Various other forms have been proposed for telpherage systems.

The porte-electric system is the name given to a system of electric carriage or transportation, by means of the successive attractions which a number of hollow helices of insulated wire exert on a hollow cylindrical core of iron.

The cylindrical car forms the movable core of a number of helical coils. As it moves through the helices it closes the circuit of an electric current through the coils which lie in advance of it, and opens the circuit of those coils through which it has just passed.

In this manner the solenoidal core advances in a line coincident with the axis of the coils, being virtually sucked through them by their magnetic attraction.

In the porte-electric system of transportation the electric motor becomes practically a mere mass of iron, as shown in Fig. 165. The system is applicable to the carriage of mails or other comparatively light articles at high speed.

The solenoidal coils are shown in Fig. 166, a

section of the track of the porte-electric system as operated on an experimental plant at Dorchester, Mass. The length of this track was 2,784 feet, the solenoidal core or car weighed about 500 pounds, and could carry about 10,000 letters. It was provided with two flanged wheels placed above and below.

The solenoidal coils, whose attractive power caused the motion of the car, embraced the track and the movable core or carrier. Each coil was formed of

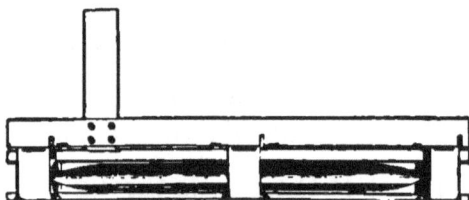

FIG. 165.—PORTE-ELECTRIC CAR.

630 turns of number 14 copper wire. A speed of about 34 miles an hour was reached.

By far the most successful system of electric propulsion of movable carriages is seen in the various systems of electric railroads which are now in such extended successful use. In such systems cars are propelled by means of electric motors.

The current that drives the motor is derived either from storage batteries placed on the cars, or from a dyuamo-electric machine especially designed for this purpose and situated at some point on the road.

The current from the generating dynamo in this latter case is led to the car along the route by means of suitable conductors, and is taken from such conductors and passed into the motors.

Systems for electric railroads may, therefore, be divided into:

FIG. 166.—PORTE-ELECTRIC TRACK.

(1.) The Independent System, where the driving current is derived from primary or secondary batteries placed on the car ; and,

(2.) The Dependent System, where the driving current is taken from conductors placed somewhere outside the car by means of sliding or rolling contacts.

The dependent system of motive power for electric railroads includes three distinct varieties.

(1.) The Underground System.

(2.) The Surface System.

(3.) The Overhead System.

In these systems, since the current is led from the generating dynamo through line wires or conductors that supply the current to the motors by means of rolling or sliding contacts, such wires or conductors are necessarily bare or uninsulated. They must, therefore, be suitably supported on insulators in order to prevent leakage, and such insulators must possess comparatively high, insulating powers.

In order to avoid the difficulties arising from bare underground wires, systems have been devised in which the conductors are entirely surrounded by insulating materials, and, either actual temporary contacts are made as the car passes, or the car takes the energy required to propel it by means of induction from the underground conductors. None of these systems, however, have come into any extended use.

In the underground system a continuous bare conductor, placed in an open-slotted conduit, supplies the driving current by means of traveling conductors or trailers placed on the car and connected with the electric motors by rolling or sliding over it.

In the surface system the wires or conductors connected to the generating dynamo, instead of being placed underground in an open-slotted conduit, are placed directly on the surface of the street or road-bed and the current taken from them by suitable contacts placed on the car.

The overhead system is the one that has come into the most extended use. In actual practice it operates by means of a continuous bare conductor suspended by suitable supports over the roadway or bed.

The current required to drive the car is taken from the overhead wire or conductor by means of a traveling wheel or roller called a trolley. In these systems the overhead wire is either arranged in a continuous metallic circuit or the ground is used for the return circuit. The latter plan is in most general use.

The electric motor is placed underneath the car on what is called a motor truck, and is geared to the axle of the car.

Most electric motors have their greatest efficiency when run at high speeds, since then their counter-electromotive force is greatest. In order to reduce the speed of the car to the limit of safety required for use in crowded cities, and yet to permit the motor to run at a high speed, some form of reduction gear is employed.

Improvements have recently been made in what are known as low-speed motors, by which fairly low speeds can be obtained, without any reduction gear whatever, with a fair degree of efficiency.

In order to regulate the speed of the motor various devices are employed, the object of which is to vary the current in the motor circuit. These devices consist essentially of rheostats, or resistances, which are introduced into or removed from the motor cir-

FIG. 167.—LIVE TROLLEY CROSSING.

cuit by the movement of a lever, or the movement of a wheel, which forms part of the circuit and moves over contact plates connected with the various resistance coils, and also of devices for readily effecting different couplings of the field coils, etc. Like all such devices, the portions handled are carefully insulated from the circuit.

In order to change the direction of rotation of the motor, and thus reverse the direction in which the car moves, various devices are employed, which depend either on the changing of the field, or reversing the direction of the current in the field or in the armature.

In order to protect the electrical apparatus from an accidental discharge of lightning through the bare conductors, some form of lightning arrester is connected with the line.

Various forms are given to the trolley arms or poles. A well-known form is known as the drop-trolley. In this form the movement of a lever connected with the trolley pole causes the trolley wheel to drop away from the line wire. A motion in the

FIG. 168.— TROLLEY CROSS-OVER.

opposite direction raises the trolley wheel upward to the proper elastic pressure.

In systems where the overhead line or conductor consists of two wires forming a continuous metallic circuit a double trolley wheel must be employed, one wheel to carry the current to the motor and the other to return it after it has passed through the motor.

A trolley cross-over is a device by means of which a trolley is enabled to pass without interruption over points where different lines cross one another. A trolley cross-over is shown in Fig. 168.

The position of the trolley arm, etc., in the case of a form of double deck car designed by the Pullman Company for street railway services, is shown

FIG. 169.—PULLMAN STREET CAR.

in Fig. 169   As will be seen from an inspection of the figure, a spiral stair case is provided at either side of the car, near the centre, to ensure ready communication with the two compartments.

EXTRACTS FROM STANDARD WORKS.

Crosby and Bell, in "The Electric Railway in Theory and Practice," * speaking of the efficiency of electric traction, on page 202 say :

Whatever may be the advantages of electric traction, whatever its convenience as a means of rapid transit, it is on its efficiency that its ultimate importance must depend. We must realize at the start that the electric motor is not a prime mover, a fundamental source of energy ; it only furnishes a very perfect and elegant means of utilizing electrical energy, already generated by some prime mover, at the point where it may be most convenient to employ it, whether that point be fixed, as in the case of stationary motors, or moving, as in the case of street railways. Advantages may be and are gained by employing motors sufficient to offset considerable losses in the necessary transmission and transformation of electrical energy, but if these losses rise above a certain amount the system must inevitably be a commercial failure.

Let us look, then, deliberately at the series of transmissions and transformations necessary in electric traction, and form as close estimates as possible of the losses, their magnitudes, and the most practicable means for reducing them to a more satisfactory figure. The first transformation of energy is from the pressure of steam generated in the

---

* "The Electric Railway in Theory and Practice," by Oscar T. Crosby and Louis Bell, Ph. D. Second edition, revised and enlarged. New York: The W. J. Johnston Company, Limited, 1893, 412 pages, 182 illustrations. Price, $2.50.

boilers to the rotary motion produced by the engine and employed in driving dynamos.

Then the mechanical energy obtained is first transferred, through the medium of shafting or belting, to the dynamo, where it is again transformed and appears as electrical current on the line. In this convenient shape it is transferred, with little loss, to the point on the line where the motor or motors may happen to be. There it undergoes another transformation in the motor back to mechanical energy, which is then transferred, through the medium of gearing of one sort or another, from the armature shaft to the car wheels.

Fortunately, the losses at several points in this somewhat complicated system of transmutations are comparatively small; and for practical purposes we may consider the losses to be substantially as follows : first, the losses in the engine and attachments ; second, those in the dynamo ; third, those on the line ; fourth, those in the motor; fifth, those in the gearing. Luckily, not all these are serious. In the art of electric traction as to-day practiced their relative magnitudes are about as follows : the most formidable are the first and last ; they are of about the same magnitude, varying enough in different cases to render it quite impossible to say off hand which is the larger. Then come the losses in the dynamo and motor, generally smaller than either of the former, and that in the motor being somewhat the larger. Finally, the loss on the line, which in many cases is the least of all. Reduction of gearing has in some cases made that source of loss relatively small, but too often at the expense of motor efficiency.

Various methods are adopted for measuring the strength of currents that pass in any circuit. The most important of these are as follows :

(1.) The voltametric method, based on the electrolytic power of the current.

(2.) The calorimetric method, based on the heating power of the current.

(3.) The galvanometric method, based on the magnetic power of the current.

(4.) The indirect method, in which the electromotive force and the resistance are first measured, and the current strength then calculated.

In the voltametric method the current strength passing is measured by means of the amount of chemical decomposition it effects in a liquid placed in an instrument called a voltameter.

Voltameters may be divided into two classes ; namely, volume voltameters and weight voltameters.

In the calorimetric method the current strength passing is measured by means of the increase in temperature it produces in a given time in a known weight of liquid placed in an instrument called a calorimeter. The heat produced by the passage of the electrical current through the conductor is proportional to the resistance of the conductor, to the square of the current passing, and to the time the current continues to pass.

In the galvanometric method the current strength passing is measured by means of the amount of deflection it produces in a magnetic needle placed in the field of the circuit.

In a galvanometer the direction in which the needle is deflected depends on the direction in which the the current flows through the deflecting circuit as well as on the position of the needle as regards such circuit; namely, whether above or below, to the right or to the left of such circuit.

In all cases the galvanometer needle tends to come to rest, under the deflecting power of the current, in a position at right angles to the direction in which the current is passing.

Various forms may be given to galvanometers; among the most important of these are the sine galvanometer, the tangent galvanometer, the ballistic galvanometer, the torsion galvanometer, the astatic galvanometer, the mirror or reflecting galvanometer and the differential galvanometer.

In the commercial distribution of electricity various devices called electric meters are employed for measuring and recording the quantity of electricity that passes in a given time through any consumption circuit.

Electric meters, though of a great variety of forms, can be grouped under the following general classes; namely:

(1.) Electro-magnetic meters, in which the current passing is measured by its magnetic effects.

(2.) Electro-chemical meters, in which the current passing is measured by the electrolytic decomposition it produces.

(3.) Electro-thermal meters, in which the current passing is measured by movements produced by the increase in temperature of a resistance through which the current passes, or by means of a difference of weight produced by the evaporation of a liquid by means of the heat generated by the current.

(4.) Electric time meters, in which no attempt is made to measure the current passing, but in which a record is kept of the number of hours during which the current flows through the consumption circuit.

The electromotive force of a source, or the difference of potential between any two points of a circuit, can be measured in a variety of ways, among the most important of which are :

(1.) By the use of galvanometers, or galvanometer-voltmeters.

(2.) By the use of electrometers, or electrometer-voltmeters.

(3.) By the method of weighing, or by balance-voltmeters.

(4.) By the indirect method in which the current strength and the resistance are first determined and the electromotive force then calculated from the formula $E = C R$.

In galvanometer-voltmeters the difference of potential is determined by the amount of the deflection of a magnetic needle produced by a current which flows through a coil of insulated wire, and which current results from the difference of potential existing between two points of the circuit whose difference of potential is to be measured. The resistance of the instrument remaining constant the value of such current depends entirely on the difference of potential of the points to which the voltmeter is connected.

The deflection of the needle may be made against the earth's field, against the field of a permanent magnet, against the action of a spring, or against the force of gravity acting on a weight.

In another form of voltmeter the strength of the current passing, and hence the difference of potential producing it, is determined by means of the heat it generates.

In the quadrant electrometer the difference of potential of a circuit is determined by the electrostatic attractions and repulsions of an easily moved metallic needle suspended between insulated metallic quadrants.

In the capillary electrometer the difference of potential is determined by the movements of a drop of acid in a capillary tube filled with mercury.

Standard voltaic cells are convenient devices for obtaining a known difference of potential which is employed to determine an unknown difference of potential by balancing or opposing it.

Some of the most frequently employed standard voltaic cells are Clark's standard cell, Raleigh's modification of Clark's standard cell, Fleming's standard cell and Lodge's standard cell.

In these cells the electromotive force is constant only when certain conditions are rigorously maintained.

The resistance of a circuit or part of a circuit can be determined in a great variety of ways. Among the most important of these are:

(1.) The method of substitution.

(2.) The comparison of the deflections of galvan·ometer needles.

(3.) The use of differential galvanometers.

(4.) The use of a Wheatstone bridge in connection with a box of resistance coils.

(5.) The indirect method ; that is, by the formula

$$R = \frac{E}{C}.$$

In measuring the resistance of a circuit by means of a Wheatstone bridge the circuit is caused to branch and flow through four arms, two of which are placed in each branch of the circuit.   A galvanometer is made to join or bridge parts of the branched circuit lying between the resistances placed in each branch.   If the resistance in one of these arms and the relative resistance in two of the remaining arms are known, the resistance of the fourth arm can be determined from the value which the remaining resistance will have when no current flows through the galvanometer.

Wire gauges are means for determining accurately the diameter of wires or other conductors.

In 1786 Galvani made an observation concerning the convulsive movements of the legs of a recently

killed frog. A few years later this observation of
Galvani's led Volta to the invention of the voltaic
pile.

A voltaic cell generally consists of two dissimilar metals, called a voltaic couple, dipping into a
liquid called an electrolyte. A difference of potential is generated by the contact of the dissimilar
metals through the agency of the electrolyte, but the
energy required to maintain a continuous flow arises
from the chemical potential energy liberated by means
of the solution of one of the metals by the electrolyte.

In a voltaic cell one of the metals of the voltaic
couple is dissolved or acted upon chemically by the
electrolyte ; the other is not acted on. The former
is generally called the positive plate and the latter
the negative plate.

In a voltaic cell the polarity is as follows : the
negative terminal of the battery is the terminal that
is connected to the plate that is dissolved or acted
on by the electrolyte ; the positive terminal is the
terminal that is connected to the other plate.

By a convention it is agreed to call that pole of an
electric source, out from which the current flows, the
positive pole of the source, and that pole into which
the current flows, the negative pole of the source.
Inasmuch as within the cell, beneath the liquid, the

current flows in the opposite direction, it is assumed for convenience that the metal most acted on is positive, and the metal least acted on is negative. This, as will be seen, makes the polarity of a voltaic cell, within the liquid, opposite to that outside the liquid.

During the action of a voltaic cell there is a tendency for the hydrogen to collect on the surface of the negative plate. This is called the polarization of the cell.

The polarization of a voltaic cell tends to decrease the current that such cell can furnish—

(1.) On account of the counter-electromotive force which such collection of gas produces, thus decreasing the effective electromotive force of the cell.

(2.) On account of the increased resistance of the cell, due to the bubbles of gas so collected.

The ill effects of polarization may be avoided—

(1.) Mechanically; by brushing off the bubbles of gas, or by permitting them to readily escape from the roughened surfaces of the plate.

(2.) Chemically; by surrounding the surface of the negative plate by a powerful oxidizing substance.

(3.) Electro-chemically; by depositing on the surface of the negative plate a coating or layer of the same metal as that of which the plate is composed.

Voltaic cells may be divided into two great classes;

namely, single-fluid cells, and double-fluid cells. In the former there is a single electrolyte, in the latter, there are two electrolytes.

In the single-fluid cell, as the name indicates, both elements are immersed in the same electrolyte. In the double-fluid cell, each element is immersed in a separate electrolyte, the fluids being kept from mixing either by means of a porous partition, or cell, or by means of their different densities.

The principal single-fluid cells are the bichromate, the Smee and the zinc-copper. The principal double-fluid cells are the Daniell, the Grove, the Bunsen, and the Leclanché.

Of the great variety of voltaic cells that have been devised, two only have survived in the struggle for existence ; namely, the gravity and the Leclanché. The former is called a closed-circuited cell, because it can remain for an indefinite time on closed circuit without polarization. The second is called an open-circuited cell, because it is only suitable for use during short intervals of time.

The gravity cell is used principally in telegraphic circuits ; the Leclanché for the circuits of annunciators, electric bells, or for similar purposes.

In the thermo cell, invented by Seebeck in 1821, two dissimilar metals, formed into a circuit by soldering their ends together, produce an electric current when one of their junctions is maintained at a different temperature from the other.

Thermo-electric cells, like voltaic cells, consist of two dissimilar substances which form a voltaic couple.

A thermo-electric battery consists of a number of thermo-electric cells connected so as to act as a single electric source.

Thermo-electric batteries are connected in series in order to add together the weak electro-motive force produced by each cell.

A photo-electric cell consists of a sheet or extended layer of selenium, so arranged that a difference of potential is produced when one of its faces is differently illumined from the other.

A selenium cell is sometimes called a selenium resistance, because its electric resistance undergoes marked variations when its faces are exposed to differences in intensity of illumination.

The following points are asserted by Van Uljanin concerning selenium cells :

(1.) That the electromotive force produced by exposure causes a current to flow from the illumined to the non-illumined electrode.

(2.) That such electromotive force immediately

appears and disappears on exposure to or removal from the light.

(3.) The sensitiveness of the cell decreases with age. This change is probably due to an allotropic modification occurring in the selenium.

(4.) When heat rays are absent the electromotive force is proportional to the intensity of the illumination.

A crystal of tourmaline acts as an electric source and produces differences of potential when its ends or poles are unequally heated.

When a liquid is forced through a capillary tube differences of potential are produced, and the tube and its moving column act as an electric source.

Currents produced by the movements of liquid through capillary tubes are called diaphragm currents. The electromotive force of such currents depends

(1.) On the material of the diaphragm.

(2.) On the nature of the liquid.

(3.) On the pressure required to force the liquid through the diaphragm.

Plants and animals act as independent sources of electricity.

Any system for the distribution of electric energy embraces the following parts; namely,

(1.) Various electric sources or batteries of electric sources.

(2.) Various electro-receptive devices.

(3.) Conductors or leads connecting the sources with the electro-receptive devices.

Among the most important systems for the distribution of electric energy are the following :

(1.) A system of distribution by means of direct or continuous currents.

(2.) A system of distribution by means of alternating currents.

(3.) A system of distribution by means of storage batteries or secondary generators.

(4.) A system of distribution by means of motor generators.

Distribution by means of direct or continuous currents, though of a variety of forms, can be arranged under the following heads:

(1.) A system of constant current distribution in which the current is distributed over a line wire or conductor in such a manner that its strength is maintained approximately constant, the electromotive force of the source being changed with changes in the resistance of the circuit.

(2.) A system of constant potential ·distribution in which a constant difference of potential is maintained on the leads or conductors to which the electro-receptive devices are connected.

In a system of constant current distribution, in which the receptive devices are connected to the line in series, each device added increases the total resistance of the circuit, and, in order to maintain a constant current strength, the electromotive force must be correspondingly increased.

In the constant potential circuit each electro-receptive device added in multiple to the mains or leads decreases the total resistance of the circuit, while each one removed therefrom increases the total resistance.

Various means are devised, whereby, in a constant current circuit, variations in the electromotive force of the source are effected in order to maintain the current strength constant despite changes in the load on the circuit. These devices are either automatic or non-automatic.

The first considerable voltaic arc taken between carbon points was obtained by Davy in 1809. A carbon voltaic arc consists of a stream of highly heated incandescent carbon vapor, which proceeds from a cavity in the positive carbon, and is directed toward the negative carbon.

Voltaic arcs may be established between metallic

electrodes. Such arcs are generally less luminous than carbon arcs, but are longer and possess a color characteristic of the volatilized metal.

During the formation of a carbon arc the carbon electrodes are consumed—the positive more rapidly than the negative. When used in an arc lamp some device must be employed to maintain them a constant distance apart. This is generally effected by placing the two carbons one above the other, and causing the upper or positive carbon to drop toward the lower or negative carbon at intervals dependent on the distance between the carbons.

Arc lamps are generally placed in the distribution circuit in series. In such cases each lamp is provided with an automatic cut-out which removes it from the circuit when it fails to properly operate, and, at the same time, establishes a by-path or short circuit past the lamp so as not to interfere with the working of the remainder of the circuit.

An arc lamp is generally provided with a hood of a conical form, which serves the double purpose of reflecting the light downward and protecting the feeding mechanism of the lamp from the weather.

The crater in the positive carbon is the main source of light in the arc lamp. When an area below the lamp is to be lighted, it is, therefore, necessary to make the upper carbon the positive carbon.

When arc lamps are required to burn for a greater number of hours than can be maintained by a single pair of carbon electrodes, an all-night lamp, containing two pairs of carbons is employed. In this device, when one pair is consumed, the current is automatically shifted to the other pair.

In the Jablochkoff candle the two carbons are placed parallel to one another and separated by kaolin or some other suitable insulating material. A Jablochkoff candle employs an alternating current so as to insure a uniformity in the consumption of the two carbons, and thus burn them both down at an even rate.

The carbon electrodes employed in arc lighting are formed of artificial carbon. A mixture of powdered coal and charcoal, mixed into a paste with tar or some other carbonizable liquid, is molded by hydraulic pressure, dried and subsequently carbonized while out of contact with the air. The carbon sticks are generally covered by an electrolytic deposit of copper.

The unsteadiness of the arc light is due mainly:

(1.) To unsteadiness in the driving power.

(2.) To imperfections in the feeding mechanism of the lamp.

(3.) To impurities in the carbons.

An incandescent lamp consists essentially:

(1.) Of the incandescing filament or conductor.

(2.) Of the inclosing glass chamber.

(3.) Of the leading-in wires.

(4.) Of the device for supporting the filament inside the glass chamber and connecting it to the leading in wires or conductors.

(5.) Of the base containing contact points to which are connected the leading-in wires.

(6.) Of a socket containing contact points to which are connected the terminals of the leads that furnish the current to the lamp.

The following steps are essential in order to prepare the carbon strips from the bamboo.

(1.) The cutting and shaping of the bamboo filament.

(2.) Carbonizing the filament while out of contact with air.

(3.) Submitting the carbonized filament to the flashing process, whereby it is rendered electrically homogeneous throughout its entire length.

(4.) Properly mounting and connecting the carbonized filament to the leading-in wires.

(5.) Driving the occluded gases out of the filament by electrically heating it while in the lamp chamber during the process of exhaustion.

(6.) The hermetical sealing of the lamp.

The exhaustion required in order to obtain a vacuum in the lamp chamber is generally commenced by the action of a mechanical pump and completed by the action of a mercury pump.

Incandescent lamps are generally connected to the mains in multiple-arc or in multiple-series. In such cases economy of distribution is best obtained when the resistance of each of the lamps is high.

Incandescent lamps are sometimes connected to the line wire or conductor in series ; in such cases the electric resistance of each lamp is low.

In multiple-connected incandescent electric lamps, the cutting out of a single lamp does not affect the other lamps in the circuit. Such circuits are protected from the presence of abnormally large currents by placing in them safety fuses or strips, which fuse and break the circuit on the passage of a current slightly less than that which the wire forming the circuit can stand without injury. * Such safety fuses, therefore, protect the wire circuits and the electro-receptive devices connected therewith from excessive currents.

The life of an incandescent lamp is rated by the number of hours during which the lamp is capable of acting as an effective source of light. The failure of a lamp to properly operate may arise either from the breaking of the filament or from a loss of transparency of the lamp chamber. This decrease of transparency may arise either from the settling of dirt on the outside of the lamp chamber, or from an accumulation of volatilized metal or carbon on the inside.

When a current of electricity flows alternately in opposite directions it is called an alternating current in contradistinction to a direct or continuous current that continually flows in one and the same direction.

In alternating currents the electromotive forces producing the currents are alternately directed in opposite directions. Their values may be represented by means of a curve.

Two or more alternating currents are said to possess the same phase when they are simultaneously similarly directed. Alternating currents are said to possess the same period when their number of alternations per second is the same. They are said to be in synchronism with each other when their electromotive forces produce currents in the same direction and for the same length of time.

When two alternating current dynamos are connected in series to the same leads they tend to drag each other into opposite phases and so produce no current. When connected to the same leads in parallel they tend to pull each other into synchronism. Series connection, or coupling of alternators, is therefore impracticable.

The following peculiarities are possessed by alternating currents:

(1.) The currents undergo regular changes in direction.

(2.) The currents undergo regular changes in strength.

(3.) The peculiarities of alternation, either in direction or in strength, during one complete alternation, that is, during one complete to-and-fro motion, are regularly repeated during any subsequent complete alternation or to-and-fro motion ; in other words, such motions are simple-periodic or simple-harmonic motions.

In some alternating currents the motions, though regularly recurring, are more complex in nature and are, therefore, complex-harmonic motions.

In alternating currents the induction of the current on itself, that is, its self-induction or its inductance, is very marked, for the current is constantly undergoing changes both in strength and in direction.

The resistance of any circuit to the passage of a current through it is called its impedance. In the case of an alternating current the impedance is equal to the sum of the ohmic resistance of the circuit and its inductance.

In any ordinary direct or continuous current circuit the current strength passing at any moment is correctly represented by the formula $C = \dfrac{E}{R}$; in which case C, is the current in ampères, E, the electromotive force in volts, and R, the resistance in ohms. In a simple periodic or alternating current the average current strength equals the average impressed electromotive force divided by the impedance. The impedance equals the square root of the sum of the squares of the inductive resistance and the ohmic resistance.

When a continuous current is passed through a conductor, as soon as the current becomes steady the current density is the same in all cross-sections of the conductor. When, however, an alternating current is passed through a conductor, the current density is greatest near the surface, and, if the rapidity of the alternation be sufficiently great, the central portions of the conductor are nearly free from current.

It is now generally believed that the old conception of a current passing through the mass of a conductor needs modification. The electric energy is now regarded as passing through the dielectric or non conducting space outside of the conductor, and as being rained down on the surface of the conductor.

The discharge of a Leyden jar partakes of the nature of rapidly alternating discharges; consequently, when passed through the primaries of induction coils such discharges produce by induction rapidly alternating currents in the secondaries of such coil.

The phenomena of the alternative path of a disruptive discharge can be explained by the oscillatory or rapidly alternating character of such discharges.

The so-called anomalous magnetization, produced by the discharge of a Leyden jar through a magnetizing spiral, is caused by the rapidly alternating character of such discharges.

Rapidly alternating currents passed through conductors produce by induction rapidly alternating currents in neighboring conductors. The effects of such induction can be avoided by plates of metal placed between the conductors, because such plates have currents produced in them and thus protect the conductors from them. This action is called screening.

A system for the distribution of alternating currents of electricity includes :

(1.) An alternating current source.

(2.) A line wire or conductor arranged in a metallic circuit. .

(3.) Transformers for changing or transforming the current and electromotive force.

(4.) Electro-receptive devices placed in the circuit of the secondary coils of the transformers.

In a system of alternating current distribution rapidly alternating currents, sent over the line, pass. through the primary coils of an instrument called a transformer placed in the same circuit, and produce in the secondary coils of such transformer rapidly alternating currents, which differ both in difference of potential and in current strength from those of the primary.

The various systems devised for the use of alternating currents can be arranged under two heads ; namely:

(1.) A system of constant potential distribution in which the primaries of the induction coils are connected in multiple to leads maintained at a constant difference of potential.

(2.) A system of constant current distribution in which the primary coils are connected in series to a metallic circuit.

The greater the electromotive force impressed upon a circuit of limited area of cross-section the smaller the amount of energy lost in transmitting a required amount of energy over such circuit.

The advantages of sending over a line wire or conductor, a current of higher potential than can be used in the distribution circuit, and subsequently lowering or lessening such potential by the use of transformers, renders the use of such alternating currents of great value in the economical distribution of electric energy, as employed in systems of incandescent lamps.

The transformers generally employed with alternating currents in connection with incandescent lamps, are of the type known as step-down transformers; that is, transformers in which a great length of comparatively thin wire, is used in the primary, and a smaller length of comparatively thick wire in the secondary.

In step-down transformers a current of great difference of potential and small current strength is transformed into a current of small difference of potential and great current strength.

When transformers are connected to leads maintained at a constant difference of potential, the current that flows through the mains is automatically regulated by reason of the self-induced counter-electromotive force set up in the primary, so as to meet the requirements of the load placed on the mains, when new electro-receptive devices are put in connection with or removed from such mains.

In actual practice this regulation is not strictly automatic, and special devices are required in order to prevent a drop of potential occurring on the mains or excessive variations in the number of electro-receptive devices placed thereon.

In systems of alternating current distribution the source consists of an alternating current dynamo-electric machine or battery of dynamo-electric machines.

The commercial alternating current dynamo-electric machines required to produce a high rate of alternation are multipolar ; that is, they possess more than a single pair of field magnet poles.

In a system of distribution by means of alternating currents, in order to protect the consumption circuit from the high potential currents sent through

the primaries, the secondaries should be highly insulated from the primaries.

The passage of an alternating current through an
incandescent electric lamp produces an alternate increase and decrease in the temperature of the lamp,
and consequently an alternate increase and decrease
in the brightness of the emitted light. Such lamps,
however, produce a steady light provided the rate
of alternation is sufficiently high.

In a system of direct or continuous current distribution by means of transformers, devices called
motor-generators are employed in order to convert
the continuous current of high potential sent over
the main line or conductor, into the character of
current required for use in the consumption circuit.
In some cases, however, devices called disjunctors
are employed to rapidly and periodically reverse the
current so as to feed the transformers, by means of
which the current is transformed to one of smaller
potential and greater current.

When two alternating current dynamos are connected in series so as to form a single source, the two
machines soon pull each other into opposite phases;
when, however, they are connected in multiple, even
though out of synchronism they soon pull each other
into synchronism.

A choking or reaction coil consists of a coil of insulated wire wound on a core of soft iron wire. Such a coil acts by its self-induction to choke off an alternating current endeavoring to pass through it with much less loss of power than if an ohmic resistance were used. The higher the rapidity of alternation the greater is the choking effect of a given coil.

A choking coil employed for the purpose of automatically regulating the intensity of the light emitted by an incandescent lamp is called a dimmer.

When the rate of alternation becomes exceedingly high, alternating currents present numerous phenomena which differ in a marked manner from those possessed by currents of but moderately high frequency.

Among such differences are :

(1.) When sent through electro-magnets the alternate attractions and repulsions of the armatures disappear, and apparently nothing remains but attraction.

(2.) Substances which act as insulators for ordinary currents act as conductors for alternating currents of enormous frequency.

(3.) Substances which act as conductors for ordinary currents will not permit alternating currents of very high frequency to pass through them.

(4.) The physiological effects of alternating currents of high frequency are much less severe than are those of but moderate frequency.

By employing alternating currents of enormously high frequency, and sending them through the primaries of peculiarly constructed induction coils, Nikola Tesla obtains a great variety of unusual and curious electric discharges. Among some of the most important of these are :

(1.) The sensitive-thread discharge.

(2.) The flaming discharge.

(3.) The streaming discharge.

(4.) The brush-and-spray discharge.

(5.) The luminous-disc-shaped discharge.

(6.) The rotating-brush discharge.

Tesla has devised a new form of electric lamp which may be termed an electric bombardment lamp, in which the light is obtained from the intense molecular bombardment of gaseous molecules under the influence of electric discharges of enormously high frequency.

Electric bombardment lamps may be rendered luminous when subjected to the electrostatic thrusts of discharges of enormous frequency when but a single pole is connected to the source of such dis-

charges, or even when no poles whatever are connected to such source.

Among the many forms of electric bombardment
lamps devised by Tesla may be mentioned the ball
incandescent electric lamp, and the straight-filament
incandescent electric lamp.

By causing the oscillatory discharges of high potential obtained from a condenser to pass through
the primaries of induction coils, Elihu Thomson has
obtained discharges from their secondaries of enormously high potential and frequency. Some of these
discharges readily pass through thirty inches of air,
at an estimated difference of potential of five hundred thousand volts.

The principles of electro-dynamic induction were
discovered by Faraday about 1831. When a conductor is caused to cut, or is cut by, the lines of magnetic force, it has differences of potential produced in
it by what is called electro-dynamic induction.

In electro-dynamic induction a motion, therefore,
is necessary either on the part of the conductor, so as
to cause it to cut the lines of magnetic force, or on
the part of the lines of force, so as to cause them to
cut the conductor; or, in other words, the magnetic
field may be either stationary or in motion, but one
or the other must be, and both may be in motion.

When the strength of a current passing through a conductor increases, the circular lines of force surrounding such conductor increase in number and expand or move outward. When the current strength decreases the circular lines of force decrease in number and contract or move inward.

Neighboring conductors, so placed as to be cut by these expanding and contracting lines of force, have differences of potential produced in them by electro-dynamic induction.

Electro-dynamic induction may be produced :

(1.) By moving a conductor through the lines of magnetic force so as to cut them.

(2.) By placing a conductor so as to be cut by the expanding or contracting lines of force.

There are four varieties of electro-dynamic induction ; namely,

(1.) Self-induction or inductance.

(2.) Mutual-induction or voltaic-current induction.

(3.) Magneto-electric induction.

(4.) Electro-magnetic induction.

Magneto-electric induction and electro-magnetic induction are sometimes called dynamo-electric induction.

In self-induction the expanding or contracting

lines of force, produced by variations in the current strength in a given circuit, are caused to cut parts of the circuit and thereby induce differences of potential therein.

Currents are produced in a circuit by self-induction both on starting and stopping a current in such circuit.

These induced currents are called extra currents-

The extra currents flow in the opposite direction to the inducing current on the completing or closing of the circuit and in the same direction on the opening or breaking of the circuit.

Since both in self-induction and in mutual induction the differences of potential are produced by expanding and contracting lines of magnetic force, and since such expansions and contractions occur only while the current strength is changing, self-induction or mutual induction is produced only while the current is undergoing a change in strength. As soon as a steady flow is established through the conductor the effects of induction cease.

In mutual-induction the expanding and contracting lines of force produced by rapidly varying the current strength in one circuit cut or pass through a neighboring circuit and induce difference of potential therein. Such differences of potential are directed in one direction as the lines of force expand or move outward from the inducing circuit, and in the opposite direction as they contract or move inward toward such circuit.

The effects of mutual-induction are utilized in those forms of electro-receptive devices termed in-- duction coils.

In an induction coil a rapidly alternating current sent through a circuit called the primary circuit, produces by induction a rapidly alternating current, in a neighboring circuit called the secondary circuit.

In magneto-electric induction a conductor is moved toward the field of a permanent magnet so as to cut the lines of force, or such field is caused to pass across the conductor by moving the magnet past the conductor.

In electro-magnetic induction a conductor is moved past an electro-magnet so as to cause the conductor to cut or to be cut by the lines of magnetic force of the magnet, or vice versâ.

Magneto-electric induction and electro-magnetic induction are, therefore, in reality one and the same variety of electro-dynamic induction and are sometimes called dynamo-electric induction.

The following general principles can be applied to all cases of dynamo-electric induction ; namely :

(1.) Any increase in the number of lines of magnetic force which pass through a circuit produces an inverse current in that circuit ; that is, a current flowing in the opposite direction to that producing the lines of force.

• (2.) Any decrease in the number of lines of force passing through a circuit produces a direct current in that circuit; that is, a current flowing in the same direction as that producing the magnetic field.

(3.) The strength of the induced current, or, more correctly, the difference of potential produced, is proportional to the rate of increase or decrease in the number of lines of force passing through the circuit.

Induction coils are forms of electro-receptive devices by means of which rapidly alternating currents, passing through a primary circuit, induce by electro-dynamic induction currents in a neighboring secondary circuit.

Induction coils are sometimes called converters or transformers. The latter term is more frequently employed.

Transformers can be divided into the two general classes of step-up transformers and step-down transformers.

In step-up transformers the length or number of turns of the primary circuit is less than the length or number of turns of the secondary circuit. When, therefore, a difference of potential is applied at the terminals of the primary circuit, a higher difference of potential is produced by induction at the terminals of the secondary circuit.

In step-down transformers the length or number of turns of the primary circuit is greater than that of the secondary circuit. When, therefore, a difference of potential is applied to the terminals of the primary circuit, a lower difference of potential is induced by dynamo-electric induction in the secondary circuit.

In the distribution of electricity by means of alternating currents a step-down transformer is generally employed.

In induction coils, as first constructed, the length of the primary was much less than that of the secondary; that is, such induction coils were step-up transformers. For this reason a step-down transformer is frequently called an inverted induction coil.

In any transformer the direction of the currents produced in the secondary circuit are as follows, namely :

(1.) Opposite in direction to that of the current in the primary circuit on making or completing such circuit ; that is, when the lines of magnetic force of such circuit are expanding.

(2.) In the same direction as the currents in the primary circuit on breaking or opening such circuit ; that is, when the lines of magnetic force of such circuit are contracting.

In any transformer the relative differences of potential in the primary and secondary circuits are proportional to the relative lengths or number of turns of such circuits.

If, for example, the length of the secondary is fifty times the length of the primary, the difference of potential induced in the secondary will be fifty times that of the difference of potential impressed on the primary.

Disregarding losses by conversion, the electric energy produced in the secondary by induction is equal in amount to the electric energy in the primary.

Representing the electric energy as the product of the current in ampères by the difference of potential in volts, $C\,E$, equals the energy of the primary, and $C'\,E'$, the energy of the secondary ; in other words, therefore, $C\,E = C'\,E'$. As much, therefore, as the current strength in the secondary is increased over that in the primary, the electromotive force in the secondary must be decreased, and vice versâ.

The principles of self-induction and mutual-induction were discovered by Professor Joseph Henry.

In the case of transformers the energy which appears in the secondary circuit is somewhat less than that expended in the primary circuit for the following reasons:

(1.) On account of the specific inductive capacity of the medium between the two circuits.

(2.) On account of hysteresis, or magnetic friction.

(3.) From loss of energy in the primary circuit from heating it.

(4.) Energy similarly expended in the secondary circuit.

In the pyro-magnetic generator electric currents are obtained by a species of dynamo-electric induction.

A dynamo-electric machine consists of any combination of parts by means of which mechanical energy is converted into electrical energy by dynamo-electric induction, by the cutting of lines of magnetic force by conductors.

A dynamo-electric machine is sometimes more broadly defined to be any machine for converting energy in the form of mechanical power into electric currents, or vice versâ, by causing conductors to move across a magnetic field or by varying a magnetic field in their neighborhood.

Dynamo-electric machines consist essentially of the following parts; namely,

(1.) A moving part called the armature, containing conductors in which differences of potential are produced. The armature is sometimes stationary.

(2.) The field magnets which produce the magnetic field.

(3.) The pole pieces which act to concentrate the magnetic field on the armature.

(4.) The commutator, by means of which the currents produced in the armature are caused to flow in one and the same direction.

(5.) The collecting brushes that rest on the commutator cylinder and carry off the current produced by the difference of potential in the armature.

The armatures are made in a great variety of forms, such as drum-armatures, ring-armatures, radial-armatures, pole-armatures.

The field magnet cores are made of massive solid iron as soft as possible; the pole pieces of the field magnets are also made of soft iron, and may be laminated in order to prevent the loss of energy from the production in them of currents called eddy currents.

The collecting brushes are formed of bundles, strips or plates of copper wire suitably soldered together, or are formed of various compositions of carbon and graphite.

The armature is made of a laminated core of soft iron in which coils of insulated wire are placed. ·

When a coil of wire is rotated in the magnetic field formed by the two opposite magnet poles, differences of potential are generated in such coil which produce currents that change their direction twice during every complete revolution.

Such currents can be made to flow in one and the same direction through a circuit external to the armature by means of devices called commutators.

Dynamo-electric machines may be divided into different classes ; namely,

(1.) Bi-polar and multi-polar machines, according to the number and disposition of the field magnets.

(2.) Into self-excited and separately-excited machines, according to the manner in which the excitation of the machine is obtained.

(3.) Into simple and compound-wound machines, according to the number and arrangement of separate circuits on the field magnet's coils.

(4.) Into various classes, according to the character of the connections between the circuit of the field magnets, the armature circuit, and the circuit external to the machine.

(5.) Into various classes, according to the character of the armature winding or the shape of the armature itself.

In the series dynamo the circuits of the field magnets and the external circuit are connected in series with the armature circuit so that the entire armature current passes through the field magnet coils.

In the shunt dynamo the field magnet coils are placed in a shunt to the armature terminals or the external circuit, so that only a portion of the current generated passes through the field magnet coils, but all the difference of potential acts at the terminals of the field circuit.

In a separately-excited dynamo the field magnet coils have no connection with the armature coils, but receive their current from a separate machine or source.

In the series-and-separately-excited dynamo the field magnets are wound with two separate coils, one of which is connected in series to the external circuit, and the other with some source external to the machine, by means of which it is separately excited.

In the series-and-shunt-wound dynamo the field magnet cores are wound with two separate coils, one of which is placed in series with the armature and the external circuit, and the other in shunt to the external circuit. Such machines are called compound-wound machines.

Electro-dynamics is that branch of electric science which treats of the action of an electric current on itself, on another current, or on a magnet. The term is used in contradistinction to electro-statics.

Ampère established the following general principles of electro-dynamics.

(1.) Parallel circuits through which electric currents are flowing in the same direction attract each other.

(2.) Parallel circuits through which electric currents are flowing in opposite directions repel each other.

(3.) Circuits placed so as to mutually intersect attract each other when the currents through them flow toward or from the point of intersection, but repel each other when the current through one of them approaches and that through the other recedes from the point of intersection.

Electro-dynamic attractions and repulsions are produced by the action of magnets on movable circuits as well as by the action of the circuits on one another. The direction of the motion produced can be determined by reference to the direction of the ampèrian currents that are assumed to cause the magnetism.

A solenoid consists of a coil of insulated wire, which acquires the properties of a magnet when traversed by an electric current.

Unlike poles of a solenoid attract each other, because the currents which produce such poles flow in the same direction in parts of the circuit lying nearest to each other. Like poles repel each other because the currents producing them flow in opposite directions in parts of the circuit lying nearest to each other.

Currents flowing in the same direction through parallel circuits attract each other because their approached lines of magnetic force extend in opposite directions; and oppositely directed lines of magnetic force attract one another.

Currents flowing in opposite directions through parallel circuits repel one another because their approached lines of magnetic force extend in the same direction; and similarly directed lines of magnetic force repel one another.

When a circuit is bent on itself so that the current in one part of the circuit flows in the opposite direction to that in the remaining part, the two parts exert no force of magnetic attraction or repulsion on external objects, because their fields neutralize one another. This expedient is adopted in the winding of resistance coils.

An electric motor consists of any combination of parts by means of which electric energy is converted into mechanical energy.

As generally constructed, electric motors convert electrical energy into mechanical energy by means of the attractions and repulsions exerted between the magnetic fields of electro-magnets, or between the fields of electro-magnets and the fields produced by currents flowing through neighboring conductors.

By the reversibility of a dynamo-electric machine is meant its ability to operate as a motor when a current of electricity is sent through its circuit.

The following analogies exist between dynamo-electric machines and electric motors :

(1.) When mechanical energy is applied to a dynamo, so as to rotate its armature, a difference of potential is produced therein ; conversely, if electric energy in the form of a current be sent through the armature and field, it will rotate and produce mechanical energy.

(2.) When mechanical energy is applied to a suitably designed dynamo, so as to rotate its armature at a constant speed, the electromotive force remains constant irrespective of the load, or irrespective of the current produced ; conversely, when electric energy is applied to a suitably designed motor, if a constant difference of potential is maintained at its terminals it will run at a constant speed approximately irrespective of the load placed on it.

The pull produced on the shaft of an electric motor by the action of the magnetic fields, or the amount of turning force the shaft exerts, is called its torque. The efficiency of a motor is the ratio between the electric energy required to turn the motor and the mechanical energy it produces.

The brushes on an electric motor require to be placed in a different position from those on a dynamo in order to avoid excessive sparking at the commutator.

In both cases a lead is given to the brushes, but in the case of the motor, in order to obtain the position of least sparking this lead is in the opposite direction to that of the dynamo. In well designed motors the amount of this lead is inconsiderable.

The reversal of the direction of motion of a motor can be obtained in various ways:

(1.) By reversing the connections of the armature.

(2.) By reversing the connections of the field magnets.

When an electric current is sent through a motor the direction of its rotation will vary with the manner in which its armature and field magnets are connected.

When a series-dynamo is used as a motor it will rotate in the opposite direction to that in which it is driven as a generator, unless the polarity of the field only is reversed, when it may be caused to rotate in the same direction as it does when driven as a generator.

When a shunt dynamo is used as a motor it will rotate in the same direction as that in which it is driven as a generator.

When a compound-wound dynamo is driven as a motor it will rotate in a direction opposite to that of its motion as a generator when its series coils are more powerful than its shunt coils, and in the same direction when its shunt coils are more powerful than its series coils.

During the rotation of the armature of a motor, an electromotive force is produced, as its wire cuts the lines of force of its field, that is oppositely directed to that of the current which produces the motion. This electromotive force is called the counter electromotive force of the motor, and acts like a resistance and opposes the passage of the driving current. As the speed of rotation increases this counter electromotive force increases, and the current required to drive the dynamo becomes less, until, when the maximum speed is reached, the driving current is very small.

When a load is placed on a motor, its speed being reduced, the counter electromotive force decreases, and a greater driving current is thus permitted to pass through it. In this way an electric motor automatically regulates the current required to drive it. An electric motor performs its maximum rate of work when this theoretical efficiency is about 50 per cent.

In the pyromagnetic motor the motion is obtained by the attraction which magnetic poles exert on an unequally heated iron disc.

A system for the electric transmission of power consists essentially of the following parts :

(1.) A line wire or conductor connecting two stations.

(2.) An electric source or battery of electric sources, generally in the form of dynamo-electric machines, placed at one of the stations for the purpose of converting mechanical energy into electrical energy.

(3.) An electro-receptive device or devices, placed in the circuit of the line wire or conductor, generally in the form of an electric motor, for the purpose of reconverting the electric energy into mechanical energy.

The circuits connecting the sources with the electro-receptive devices may be either constant-current circuits or constant-potential circuits.

The electric transmission of power possesses the following advantages over any other system, namely :

(1.) The greater distance over which power may be economically transmitted.

(2.) A greater economy.

(3.) The utilization of sources of power that would otherwise be impracticable.

(4.) A greater simplicity in the means required to connect the driving power with the driven mechanism.

The efficiency of the electric motor over that of the steam-engine will enable the electric motor to economically replace the steam-engine in all cases where electricity can be economically produced independently of the intervention of steam power.

The electric motor can advantageously replace the

steam-engine as a prime mover under the following circumstances :

(1.) Where it replaces horse power.

(2.) Wherever available water power exists.

(3.) In all places where a small amount of power is required by a sufficient number of people in any locality, and, consequently, where a single steam-engine can be employed to drive a dynamo or battery of dynamos and so supply electric power to a comparatively extended area.

The various systems of telegraphic and telephonic communication offer instances of the transmission of electric power. In both cases mechanical motions at one end of the wire are reproduced electrically at the other end of the wire or conductor.

In some systems for the transmission of electric power, the electric energy produced at one end of a line wire or conductor is utilized not only at the other end, but at intermediate points between the ends. Examples of this are seen in various telpherage systems, porte-electric systems and various systems of electric railroads.

In the telpherage system a carriage suspended from a bare line wire or conductor is propelled along it by the action of an electric motor which takes from the line the electric energy required to drive it.

In the porte-electric system of transmission a cylindrical car, in the form of a movable core, is rapidly sucked through a number of helical coils by the successive attractions which they exert on the car.

Various systems have been proposed for the electric propulsion of railway cars.

These systems may be divided into—

(1.) The independent system, where the driving current is derived from primary or secondary batteries placed on the moving car ; or

(2.) The dependent system, where the driving power is taken by means of sliding or rolling contacts from conductors placed outside the car.

The dependent system of motive power for electric railways includes :

(1.) The underground system.

(2.) The surface system.

(3.) The overhead system.

In all dependent systems, since the driving current is taken directly from the line wire or conductor by rolling or sliding contacts that move over it, such wire or conductor must necessarily be bare or uninsulated, and must, therefore, be suitably supported on conveniently placed insulators.

The overhead system, so far, is the only one which

has come into extended public use.   In it the driving current is taken from an overhead wire or conductor by means of a traveling wheel called a trolley.

INDEX.

IF YOU WISH TO KNOW

The latest and best work or works on the principles and theory of Electricity, or relating to any particular application of Electricity, The Electrical World will be pleased to promptly furnish the information, personally or by letter, free of charge. If you live in or near New York, and would like to examine any electrical books, you are cordially invited to visit the office of The Electrical World and look them over at your leisure.

Making a specialty of Electrical Books, there is no work relating directly or indirectly to Electricity that is not either published or for sale by The Johnston Company, and the manager of the Book Department keeps himself at all times familiar with the contents of every work issued on this subject at home and abroad.

Any Electrical Book in this catalogue, or any electrical book published, American or foreign, will be promptly mailed to ANY ADDRESS in the world, POSTAGE PREPAID, on receipt of the price. Address and make drafts, P. O. orders, etc., payable to

THE W. J. JOHNSTON COMPANY, Ld.,

TIMES BUILDING, NEW YORK.

ELECTRICITY AND MAGNETISM

A Series of Advanced Primers,

By Prof. EDWIN J. HOUSTON,

AUTHOR OF

"A Dictionary of Electrical Words, Terms and Phrases."

Cloth.   306 pages   116 Illustrations.   Price, 1.00.

Prof. Houston's Primers of Electricity written in 1884 enjoyed a wide circulation, not only in the United States, but in Europe, and for some time have been out of print. Owing to the great progress in electricity since that date the author has been led to prepare an entirely new series of primers, but of a more advanced character in consonance with the advanced general knowledge of electricity.

Electricians will find these primers of marked interest from their lucid explanations of principles, and the general public will in them find an easily read and agreeable introduction to a fascinating subject.

CONTENTS.

I.—Effects of Electric Charge. II.—Insulators and Conductors. III.—Effects of an Electric Discharge. IV.—Electric Sources. V.—Electro-receptive Devices. VI —Electric Current. VII.—Electric Units. VIII. —Electric Work and Power. IX.—Varieties of Electric Circuits. X.—Magnetism. XI.—Magnetic Induction. XII.—Theories of Magnetism. XIII.—Phenomena of the Earth's Magnetism. XIV,—Electro-Magnets. XV.—Electrostatic Induction. XVI.—Frictional and Influence Machines. XVII.—Atmospheric Electricity. XVIII.—Voltaic Cells. XIX.—Review, Primer of Primers.

PUBLISHED AND FOR SALE BY

THE W. J. JOHNSTON COMPANY, Ltd.,

TIMES BUILDING, NEW YORK.

The Electric Railway
IN THEORY AND PRACTICE.

By O. C. CROSBY and Dr. LOUIS BELL.

Second Edition. Revised and Enlarged.

416 Octavo Pages, 182 Illus. Price, $2.50.

This is the first SYSTEMATIC TREATISE that has been published on the ELECTRIC RAILWAY, and it is intended to cover the GENERAL PRINCIPLES OF DESIGN, CONSTRUCTION AND OPERATION.

TABLE OF CONTENTS:

APPENDICES:

Copies of this or any other Electrical Book or Books published will be promptly mailed to any address in the world, POSTAGE PREPAID, *on receipt of price.* Address

The W. J. JOHSTON CO , Ltd.,
TIMES BUILDING, NEW YORK.

Alternating Currents of Electricity:

Their Generation, Measurement, Distribution and Application.

***

By Gisbert Kapp, M.I.C.E., M.I.E.E.

With an Introduction by William Stanley, Jr.

***

Cloth. 164 Pages, 37 Illus., 2 Plates. $1.00.

This volume explains in clear, simple language the theory of alternating currents and apparatus, particular attention being paid to transformers and multi-phase currents and motors.

The treatment is entirely a practical one, the descriptions noting the various advantages and defects of different types, and the sections devoted to designing containing the practical data and instructions required by the engineer.

CONTENTS.

PUBLISHED AND FOR SALE BY

The W. J. JOHNSTON COMPANY, Ld.,

TIMES BUILDING, NEW YORK.

PRINCIPLES OF

DYNAMO ELECTRIC MACHINES

AND

Practical Directions for Designing and Constructing Dynamos,

By CARL HERING.

Sixth Thousand. 279 pages. 59 Illustrations Price, $2.50.

CONTENTS.

Review of Electrical Units and Fundamental Laws.
Fundamental Principles of Dynamos and Motors.
Magnetism and Electromagnetic Induction.
Generation of Electromotive Force in Dynamos.
Armatures.
Calculation of Armatures.
Field Magnet Frames.
Field Magnet Coils.
Regulation of Machines.
Examining Machines.
Practical Deductions from the Franklin Institute Tests of Dynamos.
The So-called "Dead Wire" on Gramme Armatures.
Explorations of Magnetic Fields Surrounding Dynamos.
Systems of Cylinder-Armature Windings.
Table of Equivalents of Units of Measurements.

Copies of this or any other electrical book or books published, will be promptly mailed to any address in the world, POSTAGE PREPAID, *on receipt of price. Address*

The W. J. JOHNSTON COMPANY, Ld.,

TIMES BUILDING, NEW YORK.

STANDARD TABLES

FOR

ELECTRIC WIREMEN

WITH INSTRUCTIONS FOR WIREMEN AND LINEMEN,
RULES FOR SAFE WIRING, DIAGRAMS OF
CIRCUITS AND USEFUL DATA.

By CHAS. M. DAVIS.

Third Edition, Thoroughly Revised and Edited by
W. D. WEAVER.

Cloth,   -   -   -   Price, $1.00.

The Third Edition of this popular work has been
THOROUGHLY REVISED and new material so extensively
added as to render it practically A NEW BOOK.

The wiring tables have been recalculated on a uniform basis and arranged in a more convenient manner
for practical use.

The object has been to produce a book for wiremen
thoroughly reliable and practical in its data and free
from verbiage and padding.

---

*Copies of this or any other electrical book or books pub-
lished, will be promptly mailed to any address in the world,
POSTAGE PREPAID, on receipt of price. Address*

The W. J. JOHNSTON COMPANY, Ld.,

TIMES BUILDING, NEW YORK.

Practical Information

—FOR—

TELEPHONISTS.

By T. D. LOCKWOOD,

Electrician American Bell Telephone Company.

12mo, 192 Pages; Cloth.

PRICE, - - - $1.00

CONTENTS.

Historical Sketch of Electricity from 600 B. C. to 1882 A. D.
Facts and Figures about the Speaking Telephone.
How to Build a Short Telegraph or Telephone Line.
The Earth and Its Relation to Telephonic Systems of Communication.
The Magneto-Telephone—What it Is, How it is Made, and How it Should be Handled.
The Blake Transmitter.
Disturbances Experienced on Telephone Lines.
The Telephone Switch-Board.
A Chronological Sketch of the Magneto-Bell, and How to Become Acquainted with it.
Telephone Transmitter Batteries.
Lightning—Its Action upon Telephone Apparatus—How to Prevent or Reduce Troubles Arising Therefrom.
The Telephone Inspector.
The Telephone Inspector—His Daily Work.
The Inspector on Detective Duty.
The Daily Routine of the Telephone Inspector.
Individual Calls for Telephone Lines.
Telephone Wires versus Electric Light Wires.
Electric Bell Construction, Part I.
Electric Bell Construction, Part II.
Housetop Lines, Pole Lines and Aërial Cables.
Anticipations of Great Discoveries and Inventions.

Copies of the above book will be sent by mail, POSTAGE PREPAID, to any address in the world, on receipt of price. Address

THE W. J. JOHNSTON CO., Ld.,

Times Building, New York.

www.ingramcontent.com/pod-product-compliance
Lightning Source LLC
Chambersburg PA
CBHW021343210326
41599CB00011B/728